Soviet Scientists Remember

Soviet Scientists Remember

Oral Histories of the Cold War Generation

Maria A. Rogacheva

LEXINGTON BOOKS
Lanham • Boulder • New York • London

Published by Lexington Books
An imprint of The Rowman & Littlefield Publishing Group, Inc.
4501 Forbes Boulevard, Suite 200, Lanham, Maryland 20706
www.rowman.com

6 Tinworth Street, London SE11 5AL, United Kingdom

British Library Cataloguing in Publication Information Available

Library of Congress Cataloging-in-Publication Data Available

ISBN: 978-1-4985-7434-1 (cloth)
ISBN: 978-1-4985-7435-8 (electronic)

To Joshua, with all my love

Contents

Acknowledgments

I want to thank members of the Chernogolovka scientific community, who were interviewed for this volume. They generously agreed to share their life stories with me and spent many hours of their time helping me better understand what it was like to live in the Soviet Union during the last four decades of its existence. Their enthusiasm for my project and their willingness to review their pasts were crucial for the successful completion of this oral history volume.

This book develops from my earlier study on the formerly closed scientific town of Chernogolovka and the complex transition of Soviet society from Stalinism into the post-Stalin era. I remain grateful to all the individuals who supported me on that project, including Jim McAdams, Semion Lyandres, Wilson Miscamble (Father Bill), Alexander Martin, Thomas Remington, and Lewis Bateman and Lucy Rhymer of Cambridge University Press. I also want to thank Donald Raleigh for his insightful comments on the early draft of this project, and my editors Eric Kuntzman, Alexandra Rallo, and Brian Hill at Lexington Books. Many thanks to the Nanovic Institute for European Studies and the Department of History at the University of Notre Dame for their generous financial support over the years.

I am especially grateful for the loving support of my family and friends. My parents, Alexander and Galina, have been a constant source of encouragement to me in all my endeavors. My brother Sergei and his wife Yuliana have kept me smiling, sending me pictures of their cat. My mother-in-law, Cheryl Bandoch, has been an important source of support. Nigel Ashford and Jennifer Prebyl have been good friends over the past couple years. Pascal has kept me delightful company, reminding me to go on walks from time to time.

My greatest thanks to my husband, Joshua, who shared his love, patience, and good sense of humor, supporting me throughout the process of writing and translating this book. He encouraged me to pursue this project in the first place and provided valuable feedback and edits on each chapter. I am grateful for his continued love, friendship, and support, and I dedicate this book to him.

Introduction

What was it like to be a Soviet citizen? What was it like to live in the first socialist society? How did Soviet citizens square the ideal of equality with the reality of mass repressions in the 1930s and other state-induced suffering? How did those and other experiences, like the Great Patriotic War (as Russians refer to World War II), impact the lives of ordinary people? What was it like to come of age during the Thaw[1] or to experience firsthand the transformation of Soviet society after the death of Joseph Stalin in 1953? What factors contributed to the idealism and optimism of the postwar Soviet youth, and how did entering adulthood during the first decade of the Cold War define young people's professional and personal life decisions? What was it like to be a citizen of one of the two global superpowers, and how much did Soviet citizens know about life behind the Iron Curtain? How did ordinary people organize their daily lives, and to what extent did they embrace the ideology of the ruling Communist Party of the Soviet Union?

One way to explore these important questions is to study the experiences of different groups in Soviet society. Despite the official claims that social classes had been abolished in the USSR, Soviet society maintained a rigid social hierarchy. At the top was a ruling political elite, followed by the scientific and cultural intelligentsia, skilled and unskilled workers, and then the peasantry. After Stalin's death, the new Soviet leader Nikita Khrushchev proclaimed the complete victory of socialism and the imminent transition to Communism. At the 1959 Party Congress, for example, he promised that in the near future the Party would be able to "fully satisfy the needs of Soviet people in food, housing, and clothes within reasonable limits." The Third Party Program, adopted two years later, pledged to complete the construction of Communist society within two decades. It defined Communism as a classless society in which all people would be equal and jointly own the means of production. In reality, however, Soviet society remained highly stratified. Some groups enjoyed a higher social status and a wide range of material privileges, including access to limited goods and services. Others survived on meager salaries and waited for years, sometimes decades, to move out of communal apartments. In this volume, I bring to light the lived experiences of one of the most influential, yet largely overlooked, groups in the late Soviet Union: scientists.

Beginning near the end of World War II, science's prestige—and import—in Soviet society exploded. It reached unprecedented proportions once the Cold War started in the late 1940s. Stalin recognized that scientists would play a leading role in building the Soviet Union into a superpower. He invested tremendous resources in large-scale technologies and training highly skilled scientists and engineers. After Stalin's death, Nikita Khrushchev doubled down on funding science. He continued to invest heavily in military-oriented scientific research. On top of that, he greatly extended the scope of the Soviet scientific enterprise by increasing funding for fundamental science.

For the new Soviet ruler, science was more than just a means to win a military conflict. Khrushchev saw science as a vehicle to solve Soviet economic and social problems, and eventually secure the USSR's economic superiority in the escalating Cold War. He recognized that the Stalinist view of science, with its focus on tight ideological control and military technologies, was detrimental to Soviet economic development. The Soviet government needed to stop stifling innovation and to allow scientists more professional autonomy in order to progress. A true Communist believer, Khrushchev viewed the scientific intelligentsia as his natural ally in the struggle to revitalize socialism, compromised by Stalin's crimes. He believed that scientists and the Soviet state had shared goals, and encouraged the input of prominent members of the scientific community.

Under Khrushchev, scientists became elite members of Soviet society. They enjoyed relative material prosperity, including privileged housing, healthcare, and higher salaries. They also gained more professional autonomy from the Communist Party, especially compared to their predecessors. Scientists managed to maintain their privileged status until the Soviet collapse in 1991. Many were grateful to the Soviet state for these outstanding professional and personal opportunities. They embraced the regime's promise to reinvigorate socialism through scientific development and innovation, and felt proud of their closeness to the Soviet state.

At the same time, scientists were among the most highly educated and critically minded individuals in the Soviet Union. Having access to information, they argued, was crucial to their professional success. Some scientists had access to foreign scientific journals and exchanged correspondences with their foreign colleagues, keeping them up to date on their scientific results.

Others were allowed to travel to Western, capitalist countries at the height of the Cold War. Many scientists also read samizdat[2] or listened to foreign radio broadcasts. The vast majority of scientists subscribed to *Novyi Mir*—the most respected literary journal in the country, which published texts by writers like Aleksandr Solzhenitsyn wrestling with the violent Soviet past.

In short, scientists occupied a highly influential, yet somewhat peculiar place in Soviet society. On the one hand, they were some of the most enlightened Soviet citizens. Their scientific curiosity often went beyond strictly professional boundaries, prompting them to ask questions about Soviet political, social, and economic matters. On the other hand, because of their prominent role the state expected their complete and unquestioning loyalty. Looking more closely at the everyday lives of the scientific intelligentsia will help us better understand the complex, changing relationship between the Soviet state and its citizens.

Unfortunately, there are few written records in the former Soviet archives that can shed light on the lived experiences of Soviet scientists, or Soviet citizens more broadly. Heavy ideological indoctrination and censorship prevented people from expressing their true thoughts either on paper or in public discussions. Soviet citizens rarely kept diaries or mentioned political issues in personal letters, since the government closely monitored private correspondences. Voicing dissenting beliefs could get someone in serious trouble, including imprisonment. While after Stalin's death it became more acceptable to criticize the Soviet government in private conversations, usually in someone's kitchen, few people dared to openly defy the official Communist ideology. Everyone knew that this could have dire consequences for their professional—and personal—lives. Many chose to remain silent when confronted with difficult decisions. They carved out spaces for their private lives, focusing on their work, family, and friends.

Oral history allows us to capture this important, otherwise inaccessible part of Soviet history. Today oral history is a powerful methodology, widely recognized across different disciplines. This was not always the case, however. While some of the first attempts to record the stories of ordinary people took place as early as the 1940s,[3] for many years historians dismissed oral interviews as less reliable than traditional, written sources. They raised questions about the accuracy of interviewees' accounts, usually recorded many years later. They also argued that memory was subjective, and that it could change depending on the social and historical environment of an interviewee. But as Donald Ritchie eloquently refutes, "oral history is as reliable or unreliable as other research sources."[4] Primary sources, which historians have long accepted as legitimate documentation, like newspaper coverage or personal diaries or letters, have their own problems. They can suffer from bias, or can be inaccurate or incomplete. Ultimately, we need to question the legitimacy and accuracy of all sources, and we do this in large part by testing them against other available materials.

Oral history is especially useful when studying authoritarian and totalitarian societies, which suffered from state-imposed restrictions on the free flow of information. Their citizens rarely kept personal diaries or papers in which they expressed their true beliefs, worried that it might get them in trouble with state authorities. Oral history helps us understand how ordinary people organized their daily lives, what was on their minds, and how they managed to fit their individual lives into the larger state project.

The oral history tradition in the West dates back to the 1960s.[5] In the United States, it was originally political historians who began using the methodology of oral history to study prominent political, military, and business personalities. By contrast, in Europe the first oral history projects were grounded in social history. Their early practitioners strove to document the experiences of working-class people, who were seen as "left behind or forgotten by the forces of progress."[6] By the late 1970s, US oral historians turned to the "bottom up" approach as well, writing about the oppressed and overlooked groups that, they maintained, were left out of traditional history narratives.[7] Most recently, Western scholars have focused on the subjectivity of memory, arguing that studying how people have internalized and interpreted what happened can be valuable in itself.[8]

As long as the USSR remained a closed society ruled by a highly authoritarian regime, it was nearly impossible for Western scholars to conduct oral history research on Soviet Russia. The only two exceptions were the Harvard Project on the Soviet Social System and the Soviet Interview Project. Carried out in 1950–1951, the Harvard Project collected materials from 764 interviews with Soviet refugees, who had escaped to Western Europe and the United States after World War II. The researchers admitted, however, that there was a problem with their sample: all interviewees were political refugees who had left their home country, mostly involuntarily.[9] A lot of them were exceedingly critical of the Soviet system. The Soviet Interview Project took place three decades later, from 1979 to 1984, and was carried out by a group of social scientists. It focused on interviewing thousands of former Soviet citizens who had immigrated to the United States in the 1970s. Most of the respondents were Jewish and had suffered from discrimination in the Soviet Union. This inevitably colored their views on everyday life in the USSR, contributing to the bias problem.[10]

At the same time, oral history research remained off-limits inside Soviet Russia. For one thing, the country continued to be isolated from the outside world even after Stalin's death. Only a select few were allowed to travel abroad, while the KGB, or the Soviet secret police, closely monitored all interactions between foreigners and Soviet citizens inside the country. For another, it was dangerous to talk about the past in a society dominated by a giant system of internal espionage operated by the KGB. It was especially perilous to discuss one's family background, since millions of people had been affected by waves of repressions, which repeatedly devastated Soviet society after the October 1917 Revolution.

The collapse of the Soviet Union in 1991 put an end to this decades-long silence. The policy of glasnost, or openness, introduced by Mikhail Gorbachev in the mid-1980s, encouraged Soviet people to reexamine their own past. As the Soviet state loosened its grip on censorship, new information about the repressions and Bolshevik crimes became widely available, flooding the public sphere. After the dissolution of the USSR and the ban on the Communist Party, many ordinary people started talking about their experiences under the Soviet dictatorship—usually for the first time in their lives. In the early 1990s, Russian and Western scholars used this opportunity to conduct a series of in-depth interviews with former Soviet citizens from various social backgrounds. These interviews became the foundation for the first edited volume based on oral history research.[11]

Naturally, some of the first oral history projects in post-Soviet Russia focused on interviewing the survivors of state atrocities, silenced under Soviet rule. In 1998, for example, Olga Litvinenko published an edited volume, which gave voice to the victims of the Soviet collectivization campaign.[12] Catherine Merridale followed suit with her riveting book on the culture of violence in twentieth-century Russia, based partially on her interviews with survivors of the state-orchestrated famine and repressions.[13] Stalin's mass repressions became a subject of separate investigations, as scholars interviewed the survivors of the Gulag, a system of labor camps in the USSR, and their families.[14] Over the last decade, historians also conducted a series of oral history projects on the Great Patriotic War, shedding light on the lives and motivations of soldiers in the Red Army, as well as the suffering and survival of ordinary Soviet citizens under Nazi occupation, especially Soviet Jews.[15]

Mass violence and state-inflicted human suffering are fundamental parts of Soviet history. But so are the stories of millions of Soviet citizens who managed to avoid state repressions and even live comfortably within Soviet society. They graduated from prestigious colleges and universities, secured decent jobs, got married, and raised their children. Some even traveled abroad for short periods of time. Why did many ordinary people continue to support the Soviet regime even after the threat of mass violence disappeared? After all, Soviet society lasted for another thirty-eight years after Stalin's death.

Oral history of Soviet society sets out to answer this question by documenting the experiences of people from all walks of life: peasants of collective farms, workers, soldiers, urban professionals, members of the intelligentsia, and even the party elite. For example, Donald Raleigh's *Russia's Sputnik*

Generation focused on the life stories of eight Soviet "baby boomers," who belonged to the professional urban class. Born in 1949–1950, his interviewees came of age "at the zenith of Soviet socialism, only to see the system crumble three decades later."[16] Paola Messana's study on Soviet communal living relied on oral interviews with the inhabitants of communal apartments to tell the story of *kommunalka*.[17] Svetlana Alexievich's works, like *Second-hand Time* and *Voices from Chernobyl*, won their author the Nobel Prize in literature in 2015.[18] Yet, we still know relatively little about the lived experiences of the scientific intelligentsia, largely because of the veil of secrecy that surrounded scientists during the Cold War and the first post-Soviet decade.

Scientists are an especially intriguing group to study, given their prominent and peculiar place in Soviet society. Who were these people and what factors influenced their formation? Why did they choose the scientific profession, and what motivated them to work for the Soviet state during the Cold War? How did scientists organize their daily lives? What were their views on the dissident movement? Did the scientific intelligentsia grow disillusioned with the Soviet project, like many members of the cultural intelligentsia?

I first started thinking about these questions back in 2009, when I traveled to Russia to conduct research for my first book on the formerly closed scientific town of Chernogolovka.[19] Located thirty-five miles to the northeast of Moscow, the town was originally established in 1956 to serve as a military-oriented testing ground for the Institute of Chemical Physics (ICP). Directed by Nikolai Semenov,[20] a prominent Soviet scientist, the ICP was one of the main participants in the Soviet nuclear program. Surprisingly, by the early 1960s, the testing ground transformed into a vibrant center for fundamental research, working at the intersection of physics and chemistry. In my book, I explain what made that possible, highlighting the exceptional atmosphere of the Thaw, which encouraged scientists to take initiative and advocate for the construction of dozens of scientific centers across the country. For the project, I initially conducted extensive archival research. But I soon discovered that to write about the private world of Soviet scientists, I would also need to carry out substantial oral history research.

From 2009 to 2013, I conducted, recorded, and transcribed over sixty in-depth interviews with twenty-eight members of the Chernogolovka scientific community, most of them in their seventies and eighties and many of them founding members of this community. During the interviews, I focused on the scientists' private lives and personal experiences before and after they moved to Chernogolovka. I also asked them about the events that were central to the formation and evolution of their identities. Most scientists agreed to meet with me at least twice. Most interviews lasted two to three hours. While I arrived at each meeting with a set questionnaire in hand, I also encouraged my interviewees to take the freedom to reflect on their experiences and discuss topics that they found important. This proved enlightening, since the scientists often raised questions that I had not previously considered. It also allowed me to establish a trusted and respectful relationship with my respondents.

Soviet Scientists Remember is the first oral history volume that focuses entirely on Soviet scientists, allowing the Cold War generation to tell their story in their own words. It brings together six first-person accounts of scientists who lived and worked in Chernogolovka, tracking their life stories from the 1930s to the 1980s. In their interviews, scientists talked about their childhoods, family backgrounds, and growing up in the Stalinist Soviet Union. They recollected their experiences during the Great Patriotic War and the hardships they endured in the immediate postwar years. Scientists explained why they chose the scientific profession and what opportunities were available to them in the heyday of the Cold War. They recounted their reactions to Stalin's death and Khrushchev's "secret speech,"[21] talked about their participation in local party organizations, and tried to make sense of their relationship with the Soviet state and the Communist Party.

Scientists also discussed their daily lives in a closed scientific town. Many emphasized that the Soviet state took good care of scientists. It provided them with free, modern housing, which was scarce in the Soviet Union. Since the vast majority of housing was owned and distributed by the state, most people waited for years, sometimes decades, to move out of *kommunalka*. The local administration tried to organize better food supplies and healthcare, and to secure better teachers to work at local schools. Fedor Dubovitskii, who was Semenov's close associate in charge of building the town, also made sure that local scientists had a robust cultural life and relative intellectual autonomy.

My respondents' professional success at a young age contributed to their positive outlook on late Soviet society. All six interviewees arrived in Chernogolovka between 1959 and 1965, during the first years of its existence. They were in their late twenties and early thirties, and were eager to accept academic positions at the burgeoning research institutes in the town.[22] Three were rapidly promoted: Georgii Manelis and Lev Vashin established their own research laboratories at the Branch of the Institute of Chemical Physics[23] in 1960, working on combustion and detonation. They would eventually become leading Soviet researchers in the field of combustion science. Vladimir Enman, an engineer, was appointed to direct the construction department of the ICP Branch. Rimma Liubovskaia and Rustem Liubovskii worked on organic superconductors, a cutting-edge research area in the Soviet and international scientific communities at the time. Oleg Efimov had an opportunity to establish his own research group, and then laboratory, at the ICP Branch, studying conductive polymers. While none of the interviewees achieved the prominence of Nikolai Semenov, the founder of Chernogolovka, they made important contributions to their respective fields of study.

The picture that emerges out of the interviews is complex. It challenges the seemingly monolithic makeup of the scientific intelligentsia. All interviewees came from different social backgrounds. They also subscribed to a wide spectrum of political opinions. Lev Vashin, for example, came from an impoverished peasant family that had benefited tremendously from the October 1917 Revolution. He remained a staunch Communist believer even after the Soviet collapse. Vladimir Enman, by contrast, was the son of a Soviet military officer who perished in the mass purges of 1937–1938. Enman remained highly suspicious of Soviet authorities and the Communist Party throughout his life. Rustem Liubovskii and Oleg Efimov were two of the few scientists in Chernogolovka who had connections to the Soviet dissident movement. Both became close friends with Kronid Liubarskii, a dissident scientist who had resided in the town from 1967 until his arrest and trial in 1972. It is enlightening to compare their interviews to the account of Georgii Manelis, who sympathized with the dissident movement but argued for its futility. Rimma Liubovskaia's interview sheds light on the challenges faced by a female scientist in the male-dominated profession.

Since I met with each interviewee at least two times, some of the same themes and stories came up more than once. To avoid repetition and occasional rambling, I have cut some parts of the interviews. I also have rearranged the order of certain passages to ensure greater clarity. For example, at times a respondent would unexpectedly change the subject, only to come back to it fifteen or thirty minutes later. Overall, though, I tried to preserve the original flow and structure of the interviews. Each chapter opens with a brief introduction, which provides background information on the interviewee and an overview of his or her life trajectory.

There are a number of common themes running through scientists' accounts. Most interviews, for example, start with their recollections of the Great Patriotic War. The war inflicted unprecedented suffering on Soviet people, including Soviet youth, whose families had to either evacuate to the east of the country or struggled to survive under Nazi occupation. Several respondents lost their fathers during the war, which put an additional strain on their families. The war also shaped the thinking of future scientists, making them more patriotic and loyal to the Soviet state. All interviewees highlighted the importance of getting a higher education, which they saw as a way out of postwar misery and poverty. Many had no financial means to attend college and were grateful to the Soviet regime for making education widely accessible. Working hard and having a university degree, they believed decades later, was crucial for their professional success.

Another recurring theme in this volume is the intimate relationship between scientists and the Soviet state. The increased state investment in scientific development created outstanding professional opportunities for hundreds of thousands of young people who flocked into the scientific profession during the Cold War. Many of my respondents enthusiastically supported the generous state funding of science, overlooking the fact that it came at the expense of the rest of the economy. Scientists criticized the economic failings of the late Soviet regime. Yet few recognized that the widespread shortages were the product of central economic planning.

Additionally, numerous Chernogolovka scientists mentioned the higher degree of professional autonomy that they enjoyed, especially in comparison to other professional groups. Half of my respondents were able to choose what scientific areas to study, focusing on fundamental, as opposed to military-oriented, research. Scientists were also one of the few groups allowed to travel abroad, although there were still numerous restrictions on their travels, as the following accounts demonstrate. The fact that scientists had more professional and intellectual freedom than the vast majority of Soviet citizens made many of them nostalgic for the Soviet state, decades after the disintegration of the USSR.

At the same time, the following chapters provide us a glimpse into what it was like to live under a highly authoritarian Soviet regime. People usually knew little about their family history, since it was dangerous to discuss one's family background in the 1930s and 1940s, when my respondents were growing up. Soviet authorities continuously purged descendants of the former nobility, the clergy, and the kulaks[24] until the mid-1950s, when Khrushchev announced that "enemy groups" no longer existed in Soviet society. Parents were afraid to tell their children about their background. They discouraged their children from asking questions that could get a family in more trouble. Vladimir Enman, whose father was shot in 1938, confessed that he had concealed his father's repression even from his wife until 1956, when the official rehabilitation papers arrived. He was afraid that his family's tragic history might disqualify him from attending college, or might even get him arrested and sent to the Gulag. Most of my respondents knew little about their grandparents or their family's social status before the October 1917 Revolution.

Fear remained pervasive in Soviet society even after Stalin's death. Although the state no longer resorted to mass violence under Khrushchev and Brezhnev, it continued to persecute people who had the courage to openly challenge the decisions of the Communist Party. As the interviews with Rustem Liubovskii and Oleg Efimov demonstrate, the line between public and private dissent could become blurred at times, making any nonconforming citizen a potential target of KGB persecution. It is perplexing that even decades after the Soviet collapse, many of my respondents were reluctant to discuss their views on the Gulag, which they claimed they had known little about. Some were also hesitant to talk about their membership in the Communist Party or their relationship with the Soviet dissident movement. The overall picture that emerges out of the interviews is anything but black-and-white. It was not unusual, for example, for an individual to sympathize with the dissident movement or read samizdat, and continue to be a staunch supporter of the regime during the late Soviet era. One could be critical of some aspects of official Soviet policies, but enthusiastically support others.

Perhaps one of the most surprising—yet in hindsight logical—discoveries I made, while working on this volume, was that all of my interviewees were the products of Soviet society. Many internalized its values, such as collectivism, equality, patriotism, and hard work. Consider the language in each interview: "we" instead of "I," and "us" instead of "me." From an early age, my respondents were raised to believe that the individual was subordinate to the collective. The individual's interests, they thought, were secondary to the goals of the state. Soviet scientists learned collectivist thinking in kindergarten and school, as well as in various Communist youth organizations, in which millions of Soviet children were required to enroll. On account of this ideological training, many of my respondents grew up with a strong aversion to individualism, which Soviet propaganda portrayed as a reprehensible quality.

Most of my interviewees could not imagine living in a nonsocialist country. This was partially because living under socialism was their only frame of reference. Since most Soviet citizens were not allowed to travel abroad, they had no first-hand experience with how people lived in the West, or beyond Soviet borders in general. Those scientists, who received permission to travel to scientific conferences, usually spent all their time at conference sites or in scientific labs. Before their trips, they got detailed instructions from the Communist Party about how they should behave abroad. Scientists' limited interactions with their foreign colleagues didn't give them an opportunity to learn what it was like to live under capitalism.

At the same time, heavy censorship and restrictions on the free flow of information inside the USSR prevented Soviet citizens from learning about life behind the Iron Curtain. Soviet newspapers and television painted a one-sided, often deceptive picture of domestic and international affairs. Yet scientists didn't fully trust the broadcasts of foreign radio stations, such as the Voice of America, either. Curiously, even though my interviewees criticized the economic stagnation and shortages during the late Soviet era, few attributed them to the lack of economic freedom or the absence of a free market. They recognized the need for reforms, yet many felt grateful to the Soviet state on account of everything it provided them.

Equality was another value embraced by each of the six interviewees in this volume. Scientists remembered feeling as if they were "all equal" when they first moved to Chernogolovka. The local administration promoted this feeling by providing all newly arriving employees with free, modern housing, generous health benefits, and good educational opportunities for their children. Scientists occupied different professional positions, of course, and there was a certain hierarchy inside research institutes. Still, one could rise to be a head of laboratory or department, as long as she or he was ambitious and had good scientific results.

Ironically, many of my interviewees failed to acknowledge that their relative equality could only exist within the boundaries of a closed, privileged community. After the drastic decline of state funding in the 1990s, scientists sadly discovered that local research institutes struggled to function properly. This largely explains scientists' nostalgia for the Soviet era, as well as their ambivalent views on Gorbachev's perestroika in the late 1980s and Russia's attempt to transition to a market economy in the 1990s.

Talking to me two decades after the Soviet collapse, scientists reminisced about their lives in the country that no longer existed. They were finally free to openly talk about their childhood and family background, their work, their beliefs, and their political concerns. Their oral histories tell us captivating stories of early struggles and perseverance. They are also full of nostalgic recollections of scientists' professional and personal successes. Their stories help us better understand why most scientists supported the Soviet state, as well as the relative stability of late Soviet society more broadly.

NOTES

1. A brief period of ideological relaxation during the Khrushchev era (1953–1964).
2. Samizdat (lit. "self-publishing") was underground literature, banned by the Soviet government, but secretly produced and distributed in the USSR after Stalin's death.
3. Columbia University established the first Oral History Research Office in 1948.
4. Donald A. Ritchie, *Doing Oral History*. Third Edition (New York: Oxford University Press, 2015), p. 9.
5. The Oral History Association was founded in 1967.
6. Ritchie, *Doing Oral History*, p. 6; Lynne Abrams, *Oral History Theory* (London; New York: Routledge, 2010), p. 5.
7. Donald A. Ritchie, ed., *The Oxford Handbook of Oral History* (New York: Oxford University Press, 2012), p. 4.
8. Abrams, *Oral History Theory*, p. 7. See also Ritchie, ed., *The Oxford Handbook*, p. 80.
9. Alex Inkeles and Raymond A. Bauer, *The Soviet Citizen: Daily Life in a Totalitarian Society* (New York: Atheneum, 1968), p. 5.
10. James R. Millar, *Politics, Work, and Daily Life in the USSR: A Survey of Former Soviet Citizens* (Cambridge: Cambridge University Press, 1987), pp. 18–22.
11. Daniel Bertaux, Paul Thompson, and Anna Rotkirch, eds., *Living Through the Soviet System* (New Brunswick and London: Transaction Publishers, 2005).

12. Olga Litvinenko and James Riordan, eds., *Memories of the Dispossessed: Descendants of Kulak Families Tell Their Stories* (Nottingham: Bramcote Press, 1998).

13. Catherine Merridale, *Night of Stone: Death and Memory in Twentieth-Century Russia* (London: Viking, 2000).

14. Orlando Figes, *The Whisperers: Private Life in Stalin's Russia* (New York: Metropolitan Books, 2007); Jehanne M. Gheith and Katherine R. Jolluck, *Gulag Voices: Oral Histories of Soviet Incarceration and Exile* (New York: Palgrave Macmillan, 2011).

15. Catherine Merridale, *Ivan's War: Life and Death in the Red Army, 1939–1945* (New York: Metropolitan Books, 2006); Laurie R. Cohen, *Smolensk under the Nazis: Everyday Life in Occupied Russia* (Rochester, NY: Rochester University Press, 2013); and Anika Walke, *Pioneers and Partisans: An Oral History of Nazi Genocide in Belorussia* (New York: Oxford University Press, 2015).

16. Donald J. Raleigh, *Russia's Sputnik Generation: Soviet Baby Boomers Talk about Their Lives* (Bloomington; Indianapolis: Indiana University Press, 2006), p. 5. See also Donald J. Raleigh, *Soviet Baby Boomers: An Oral History of Russia's Cold War Generation* (New York: Oxford University Press, 2011).

17. Paola Messana, *Soviet Communal Living: An Oral History of the Kommunalka* (New York: Palgrave Macmillan, 2011).

18. Svetlana Alexievich, *Secondhand Time: The Last of the Soviets* (New York: Random House, 2016). See also Svetlana Alexievich, *The Unwomanly Face of War: An Oral History of Women in World War II*. Translated by Richard Pevear and Larissa Volokhonsky (New York: Random House, 2017); Svetlana Alexievich and Keith Gessen, *Voices from Chernobyl* (Normal: Dalkey Archive Press, 2005).

19. Maria A. Rogacheva, *The Private World of Soviet Scientists from Stalin to Gorbachev* (Cambridge: Cambridge University Press, 2017). Chernogolovka remained closed to foreigners until the late 1980s.

20. Nikolai Semenov was a prominent Soviet chemist and a member of the Soviet Academy of Sciences. He was the first Soviet scientist to receive the Nobel Prize in chemistry in 1956 for his discovery of chain reactions. In 1961, Semenov was also elected a candidate member of the Central Committee of the Communist Party—a position he occupied until 1967.

21. Nikita Khrushchev, the First Secretary of the Communist Party of the Soviet Union, publicly denounced Stalin's crimes at the Twentieth Party Congress on February 25, 1956.

22. The other two research institutes in Chernogolovka in the mid-1960s were the Institute of Solid State Physics and the Landau Institute of Theoretical Physics, founded in 1963 and 1965 respectively.

23. The Branch of the Institute of Chemical Physics, or the ICP Branch, was originally the main research institute in Chernogolovka. In 1991, the ICP Branch was reorganized into the Institute of Chemical Physics of the Russian Academy of Sciences in Chernogolovka. It was renamed as the Institute of Problems of Chemical Physics in 1997.

24. Kulaks were the most successful and well-off peasants. They became one the main targets of the collectivization campaign.

Chapter 1

"My Life, My Family's Life Is the Typical Story of Millions of Soviet Families"

Rimma Nikolaevna Liubovskaia (née Stepanova)

I first met Rimma Nikolaevna Liubovskaia (b. 1937) in the kitchen of her three-room apartment in Chernogolovka in early fall 2010. At the time, I was searching for private archives of Soviet scientists and visited Rustem Bronislavovich Liubovskii, a physicist I had interviewed nine months earlier, to look through some of his personal materials. After Rustem Bronislavovich and I spent several hours going through fragile, yellowed, and dusty papers, Rimma Nikolaevna invited us to the kitchen, where she already had a table set up. I soon discovered that Rimma Nikolaevna was an accomplished scientist herself. A doctor of chemical sciences, she headed a scientific lab at the Institute of Problems of Chemical Physics in Chernogolovka and had dozens of peer-reviewed publications. Somewhere in the middle of our conversation, I realized that this warm and hospitable woman had her own astounding story to tell. I asked Rimma Nikolaevna if she would be willing to give me a recorded interview, and she readily agreed.

Our first interview took place on October 3, 2010. We met again on February 16 and 27, 2011. Rimma Nikolaevna turned out to be an excellent narrator—engaging, thorough, and passionate. Her story sheds light on the origins of the late Soviet scientific intelligentsia and the exceptional opportunities available to postwar youth in the early Cold War years. Her account gives us a glimpse into everyday life in a closed scientific town and the challenges faced by female scientists in a mostly male-dominated profession in the USSR. Moreover, Liubovskaia's interview presents us with a fresh perspective on the relationship between scientists and the post-Stalin regime.

Rimma Nikolaevna Liubovskaia (née Stepanova) was born in Aleksandrovskoe, Stavropol'skii krai, on December 21, 1937, at the very height of Stalin's mass repressions. While her parents were not targeted in the terror, her maternal grandfather, a priest, was arrested and sent to the Gulag in 1937, even before Rimma Nikolaevna was born. This tragic event was never discussed in the family, yet Liubovskaia sensed, already as a child, her grandmother's disdain for Stalin. The Great Patriotic War further undermined the well-being and happiness of Liubovskaia's family. Her father was drafted into the military in 1942 and killed a few years later. Her mother, brother, and Rimma Nikolaevna ended up living on the territory occupied by the Nazis for several long months. Like millions of Soviet children, Liubovksaia starved badly during the war. After the war, her family's suffering became so intense that her mother decided to leave Stavropol'skii krai. Running away from the famine, the family moved to Central Asia in 1947. For a while, they lived in Quvasoy, a city on the border of present-day Uzbekistan and Kyrgyzstan, where Liubovskaia's mother strained her health working at an asbestos factory. They later relocated to Kyzyl-Kiya, a miner's town in the Kirghiz Soviet Socialist Republic.

Several times during our conversation, Rimma Nikolaevna jokingly called herself "an exemplary student" and "a true pioneer." In her early seventies, Liubovskaia was still convinced that getting education, especially a higher education, opened up myriad opportunities to her. It allowed her to find a way out of postwar misery and desolation and escape the extreme poverty that her family had endured for over a decade. Getting a higher technical education provided Liubovskaia with a certain degree of physical mobility: she left Kyzyl-Kiya for Moscow in 1954, moved to Irkutsk in Siberia five years later, and went back to Moscow in 1961 to apply for graduate school. Studying at the Mendeleev Institute of Chemical Technology also helped Liubovskaia join the growing ranks of the Soviet scientific elite. Remarkably, Rimma Nikolaevna didn't know she would become a scientist until after she graduated from the Mendeleev Institute in 1959. Yet, like many of her peers, Liubovskaia became fascinated with the exceptional opportunities that the profession of scientist offered in the late 1950s and early 1960s. She decided to continue her studies, defended her PhD dissertation at the Institute of Organic Chemistry in 1965, and immediately found employment in Chernogolovka.

When Rimma Nikolaevna moved to Chernogolovka, it was a closed scientific community, access to which was restricted to foreigners. She started working on the synthesis of organic superconductors at the Department of Kinetics and Catalysis at the ICP Branch. Combining work and family life, however, turned out to be challenging. After Liubovskaia got married and had her first child in 1967, she noticed a visible decrease in support for her research on the part of her mostly male colleagues. It took her almost two decades to prove that her research was of high quality. Ironically, it was the opening of the Soviet Union to the West during perestroika that helped Liubovskaia receive professional recognition first at home and then abroad.

At the same time, living in a closed scientific town provided Rimma Nikolaevna with significant material privileges. She received a separate one-room apartment in late 1965, as soon as she defended her PhD dissertation. In 1967, after their first child was born, Liubovskaia's family moved into a three-room apartment. While Chernogolovka could not boast superior food supplies in the 1960s or 1970s, the town had free daycare, high-quality schools, and better health care—all conducive to raising a family.

Liubovskaia's relationship with the Soviet state was complicated. On the one hand, she was obviously a product of the Soviet system. Her constant use of plural pronouns like "we" and "us," instead of "I" and "me," throughout the interview signaled that she identified with the Soviet collective. Rimma Nikolaevna also clearly valued the educational and professional opportunities presented to her by the post-Stalin regime. On the other hand, Liubovskaia's interview indicates that she started noticing economic and political problems in the Soviet system at least since the mid-1960s. Her husband's friendship with the dissident Kronid Liubarskii in the late 1960s and early 1970s, reading samizdat, as well as Liubarskii's arrest in 1972, made Rimma Nikolaevna even more critical of the late Soviet regime. By the late 1970s and 1980s, she had fully developed political views, which differed significantly from official Soviet propaganda. Yet she never considered herself a dissident, prioritizing her work and family above all else.

I was born on December 21, 1937, in Aleksandrovskoe village, Stavropol'skii krai. The date when I was born was also Stalin's birthday, and all my life I was constantly reminded of that. For example, when I was in third grade, one day a teacher asked me to leave the classroom. It surprised me, since I was a straight "A" student and could not think of any previous misbehavior on my part, but I went out. Several days later, on December 21, I received a fountain pen as a present from my classmates. It was my first fountain pen, and I didn't even know how to use it yet. It turned out that the teacher had collected some money from my classmates and bought two pens: one for me and one for another student, who was also born on Stalin's birthday. I still remember that pen. It even made me feel a bit proud—the fact that I owned a pen like that! (laughs)

Going back to Aleksandrovskoe, I was born there in December 1937. Aleksandrovskoe was a fertile and prosperous region. But the outbreak of the Great Patriotic War changed all that. My dad was drafted into the army. My mom, brother, and I lived under the German occupation for several months.[1] There was heavy fighting in our region: the Germans wanted to capture the oil fields in the Caucasus, and they occupied our village on their way there. My dad was killed when I was still little. As far as I remember, he worked as a salesman before the war. My mom was a teacher by training, but she stayed home with my brother and me. I also had a wonderful grandmother.

She was a priest's widow. Her husband was arrested in 1937, after someone had denounced him. He perished in prison. My grandfather was arrested even before I was born. So all I know about him comes from some of the conversations I overheard at home. I didn't understand much at the time. Only later did I start putting things together.

After the Germans occupied our village, we moved to the very edge of Aleksandrovskoe. It was the winter of 1942–1943. What was my most vivid memory of the occupation? One day they found a dead German officer in front of our house. It was announced that one hundred people from our neighborhood would be shot in retaliation. My mom, brother, and I were among them. Everyone was terrified, of course, but then the Germans somehow found out that the officer had died a natural death, and they let us go. Before that, though, I remember they brought all of us to the edge of the village where there was a statue of Lenin. The Germans dug the statue out and made a large pit in its place. They said they were going to bury people there. My mom kept telling me to keep my mouth open. I still don't know why. I also remember the bombing of our village. I remember that, when the German troops began to withdraw, they were burning down houses in Aleksandrovskoe. My family was hiding in a slit trench. I guess these were very powerful memories, because they have stayed with me ever since.

Eventually, the Germans left, and in 1944, I went to school. Our school was located in the former cadet college. It was an enormous building with huge windows, perhaps the most massive building in the village. Shortly afterwards, a famine broke out. There was a serious drought in the Northern Caucasus in 1946, and we starved badly.[2] My mom didn't work, and I guess we had no food at home, because at some point I went to school absolutely starving. I remember there was a mill in our village, and sometimes they would bring mill cake (*zhmykh*) to school, made of sunflower seed skin. Our teachers allowed us to eat it right in class. It was a difficult and hungry time. Perhaps other families did slightly better, but I had no father and my mother didn't work. At some point, all our supplies came to an end.

How did your family survive then?

I don't know, but we lived terribly. We had a cow, and it saved us for a while. We also had a small plot of land, which served three or four houses. But then in 1947 foot-and-mouth disease broke out, and we lost our only cow, our only provider. It became utterly unbearable after that. My mom had brothers who lived in Central Asia, and somehow she decided that we should go to Central Asia, that it would be better there. So in 1947, we moved to Central Asia. It was only several years after the war. We traveled there in a day coach (*obshchii vagon*). It is hard to imagine now what it looked like. We first took a train to Mineral'nye Vody. Then we had to travel to Baku to take a steamboat from there. I traveled with my mom: my brother had left earlier and my grandmother went to live with her other daughter. I remember that trip very well, with all its striking details. On the way there, my mom met a woman who also was traveling with a daughter, and the four of us tried to stay together. I remember that when we reached Baku and went from a train station to a marine passenger terminal, we had to go through mandatory sanitary treatment. We had to undress and go to a bath-house (*bania*). They washed and treated our clothes at high temperature and then gave it back to us.

And what did a day coach look like?

It was quite awful! It was one long carriage, without any benches or beds inside. When we were boarding the train, there was such a scramble. My mom had a small suitcase with her, which looked like a chest. Someone knocked it out of her hands and threw it on the ground so that people could use it as a step to board the train. The day coach was extremely overcrowded, and our little chest was our only refuge in this chaos. I remember I was sitting on the chest during the trip and my mom was sitting by my side. There were hundreds of people all around us. We were starving quite badly, but I guess my mom had been preparing for the road. After we boarded the train, she opened a bag that had little pierogis with kidney beans. Sixty years have passed, maybe more, but I still remember the smell of these pierogis.

Can you tell me a bit more about your family and your grandmother?

She was my maternal grandmother. She married a priest, but it is a separate story. My son Pavel is trying to find out where my grandfather was killed. We still don't know. I managed to find my father's grave recently, but we still don't know about my grandfather. You see, my life, my family's life is the typical story of millions of Soviet families. Everything that we had to endure—the occupation, the war. When my dad was drafted into the military in late 1942, he was too old to be drafted: he was almost fifty years old. But they drafted him anyways. There was heavy fighting near Kerch; it was a complete slaughterhouse. Many people died there, both soldiers and civilians, and my dad was one of them. Yet for many years after the war, we didn't know that he had been killed, because one of the survivors told my mother that he had been taken prisoner. And all her life she kept looking for my father, waiting for him to come back. Only recently did my family find out that he had been killed in February 1944.

As for my grandfather, he worked as a priest in the 1930s, but had to pay a high price for this. After he was arrested, his six children were banned from pursuing higher education. My mom managed to get some education, so she could work as a teacher. But the rest of the children were not allowed to attend college. My mom told me that they lived in extreme poverty. She didn't say much else. It was typical for that era: parents were scared for their children and, trying to protect them, didn't really say anything to them. I should tell you, though, that I was an exemplary student in school, a true pioneer.[3] Now people are making jokes about it, but things were different back then. After we moved to Central Asia, I was always at the forefront of different groups and organizations. Perhaps you have seen some old Soviet movies depicting pioneer rallies. During these rallies, someone always steps forward and screams: "Thank you to Comrade Stalin for our happy childhood!" I was almost always that person. And indeed for a long time we thought that we had a happy childhood. Even though I didn't have a pair of shoes that I could wear on stage to make my little appearance! My mom had to borrow shoes from our neighbor so that I could walk on stage and scream this slogan, and then she would return the shoes. It was all part of my world growing up. Only now do I understand that it was also part of the propaganda, that when they keep beating you on the head with something, you start believing it. Yet my family never talked about it.

I knew, though, that my grandmother was really waiting for Stalin's death. She was a religious person, even though she never insisted that I go to church. She would pray in the evenings. It was a dangerous time: one was scared to even step on a newspaper if it had Stalin's portrait somewhere in it. But my grandmother always referred to him as "Satan." I was a child, but I understood that I could not repeat this anywhere. I guess children are sensitive this way. And my grandmother dreamed of living long enough to see Stalin's death. It appeared to her that something would change then. In 1958, when I was still in college, she and her sister came to visit me in Moscow. My grandmother wanted to go to the mausoleum where Lenin and Stalin were at the time displayed. She wanted to make sure that Stalin was indeed dead. It is curious, but I remember very little from that visit to the mausoleum: some red light inside and that was it. I guess it didn't make any impression on me. I was already an adult, and this type of thing didn't really interest me.

Do you remember whether and when your parents started discussing such things more openly at home?

No. I left for Moscow quite early. In 1954 I was already in Moscow. I remember the Twentieth Party Congress, during which Khrushchev denounced Stalin. We had Komsomol leaders who were invited to closed meetings after the Congress. Rumors about the Twentieth Congress began to spread almost immediately. It was both striking and marvelous. But I don't remember any conversations at home prior to that. It all began after I moved to Moscow.

We will come back to this, but for now let's talk a bit more about your family and your childhood. So in 1947 you and your mother moved to Central Asia.

Yes, in 1947 my mom and I moved to Central Asia. I was in third grade then, and I didn't really know where Central Asia was or where Quvasoy,[4] a city where we were going to, was located. So my friend and I spent a lot of time looking at the map, and we finally found it. It was far away. I don't remember how long it took us to get there, but I know that when I took a train from Central Asia to Moscow—when I was going to apply for college in Moscow—it took me seven days to get there. My life was full of adventures like this, but my optimism always saved me. For example, I always knew that, despite all hardships, one day I would go study in Moscow. So I was indeed a true pioneer! (laughs)

When we arrived in Quvasoy, I remember it was very hot there. So I asked someone: "Where can I find water?" They showed me an aryk, which is an irrigation ditch common in Central Asia, and said: "Right there." I looked at it, but the water in the ditch was not clean at all. That was my first impression of Central Asia. Quvasoy is a city right on the border between present-day Uzbekistan and Kyrgyzstan. When we arrived there, we found out that both of my mom's brothers were still in the army in the Far East. Their wives were not really glad to see us. We lived with one of them for a little while, but then my mom found a job at an asbestos factory. It was a terrible job, but the factory gave us a room where we could live. The room was absolutely empty, and we had no furniture either. So they gave my mother a sheet of asbestos, which we used as our bed. We would put some blankets on it and sleep like that.

As a chemist, I understand now how dangerous it was for our health. Back then, however, asbestos slates were used for roofing (they don't use this material any more). As I said, it was a horrible job: men had to bend these asbestos slates around metal reels and cut them up and women had to carry them. My mother's back and shoulders were all bent because of that job. But she had no choice: she had to feed her family. My brother also began working so that he could get bread rations. But we starved even worse there, and no one really helped us. This was not because they were bad people: everyone lived like this. I would walk around searching for dried apricot pits, and we ate them. At some point, my mom also decided that we should bake corn flatbread and sell it. So while she was at work, I had to walk with some women across rivers in the mountains to buy corn flour and then carry it back.

When did you find time to go to school then?

Well, of course, I had time to go to school! I became a straight "A" student right away. At some point, though, I remember I went to school hungry for three days in a row. I had nothing to eat for three days. One day my teacher, Klavdiia Andreevna, came up to me and gave me a little package. I opened it and saw two pieces of bread inside. I could barely wait until the end of class. Since I was a good student, I could not really eat it in class. I waited until the class ended and then ate the bread. When I came home, I asked my mom if she had asked my teacher to give it to me. But she said "No." Neither did my brother. It turned out that it was entirely my teacher's initiative. I guess she saw my condition and brought some bread for me. I felt ashamed: I didn't even thank her. I remember waiting for her after classes several days in a row to thank her. There were people like her back in the day, too.

Then my mom transferred to work at a different factory. There was a cement factory in Quvasoy, which delivered gley from Kyzyl-Kiya in Kyrgyzstan. So we moved to Kyrgyzstan and my mom started working at a factory that supplied gley to Quvasoy. Kyzyl-Kiya was a miner's town, situated among the mountains. There were two schools there: middle school and high school. The middle school was only about three kilometers away from our apartment building, so it was not too far for me to walk. After I graduated to high school, though, I had to walk six kilometers each way to school and back every day. When I think about it now, I understand that it was not a safe place to live. There were many deported people there: the Crimean Tatars, Chechens, and Meskhetian Turks. There were high levels of crime; sometimes people were murdered. Yet I tried to stay engaged in all kinds of extracurricular activities. I participated in various athletic groups (*sektsii*) and went to different intellectual contests (*olimpiady*).

I am surprised to hear that there were athletic groups and intellectual contests there.

Most of them were at school. There was also the Palace of Culture there, but most of it was at school. For example, I remember my teacher of German language, Kamila Antonovna Vishnevskaia. She was a repatriated Pole. She organized a dance class for students and taught us how to dance Cracovienne.[5] It was quite basic, but she did her best to teach us something. She invited us to her house and showed us different art albums. For example, she first told me about the Tretiakov Gallery in Moscow. We had a good geography teacher and a good math teacher, too. They all tried to engage us, to teach us something that we would not learn otherwise. Of course, I also read avidly. I read anything I could find. We didn't have books at home, so I borrowed books from a school library. I read without any guidance, without any plan. I just read. We had a small radio set at home, and since we didn't have any lamps I would use its little red light to read. Only when I was in high school did I get something like a desk.

Did you have a one-room apartment in Kyzyl-Kiya?

Our entire apartment consisted of one room; we also had a shared porch and a stove. It was a scary place to live. I was one of the few children in our apartment building, and I had to get around everywhere by myself. I had some school friends, but they all lived far away. I had to walk alone across mountains to visit them.

I finished school in Kyzyl-Kiya with a gold medal, which allowed me to apply to any university in Moscow.[6] But it was not that easy! I finished school in 1954 with the highest grades, and I was told that I needed to get an "A" on my final essay (*sochinenie*) to be awarded a gold medal. They sent my essay and the essay of another student to Osh, which was a regional center. But people who graded our work in Osh gave us both a "C," which meant that we would not receive a gold medal. However, after that they refused to issue our high school diploma to us for a month. It turned out later that they had sent our paperwork to Frunze.[7] Since I needed my diploma to apply for college, I kept coming to school every day, walking six kilometers there and back, asking the school administration to return my paperwork. I could not apply anywhere without a high school diploma. Then, all of a sudden, they told me that they had received a response from Frunze and that both of our essays got an "A." At the end of the day, we were awarded our gold medals. These were the first gold medals at that school. However, by the time I got my diploma, the entrance exams in most universities had already ended. Despite this, I immediately boarded a train to Moscow. The trip to Moscow took me seven days, and since I didn't have much money, I could only afford a seat on the so-called third shelf.[8] I could not even sit up straight there.

Why did you go to Moscow? What were you going to do there?

I wanted to apply to the Mendeleev Institute of Chemical Technology.

How did you learn about it?

I learned about it from my chemistry teacher in ninth or tenth grade. She said that there are institutes in Moscow that rarely admit girls. I immediately decided that I needed to study there, so I went to Moscow. When I arrived, I had a big suitcase with me, which had a pillow, some books, and a thin jacket that I had made out of my old coat. I remember I stepped off the train at the Kurskii railway station and started smiling, almost uncontrollably. I could not believe I was in Moscow. I had to find a small sanatorium, called "Red Rose," where some of my uncle's distant relatives lived. They allowed me to stay with them for a little while.

The next morning I went to the Mendeleev Institute to schedule my interview. But when I arrived there, I found out that the admissions period had already ended. They told me that I had to go find a committee in charge of gold medalists, located somewhere in the center of Moscow. I don't remember how, but I managed to find that committee. They didn't want to recommend me for an interview at the Mendeleev Institute, though. "Why did you come here instead of applying for a college in Frunze or Tashkent?" they asked me. My response was: "You should have written in Soviet newspapers that students from Central Asia cannot apply to Moscow colleges then." Eventually, they set up an interview for me at the Mendeleev Institute.

Several days later, I had my interview. All interviews were carried out at the office of the Institute's rector. There were a lot of students there waiting; most of them came with their parents. One girl came with her father, who was a decorated army general. I could not help but compare her with myself, a naive girl from some remote place in Central Asia. My chances were certainly not high. At the end of the day, they simply decided to give me a chance. One of my interviewees, Professor Lebedev, began to ask me questions. He asked me why I decided to apply to this particular institute. He asked me about my background and about my father. I said he was killed at war. After that Lebedev asked me what department I had applied for. When I said I had applied for "organic chemistry," he recommended that I enroll at the Fuel Department instead, since they paid their students higher stipends. This is how I found out that I had been admitted to the institute. But did I believe it? Of course not! I stayed at "Red Rose" for another month, sleeping on the floor the entire time. And every day I would go to the Mendeleev Institute, hoping that they posted their list of newly admitted students. I needed to see my name among them. At the very end of August 1954, they finally put up the list, and I realized, with relief, that my name was on it. This is how I became a student.

Did you not go home that summer?

No. I went home the following summer. Since they admitted me late, at first I didn't get a room in the dormitory. I ended up renting "a corner" in the basement of some old house. It was not even a room, just an old sofa where I could sleep. My stipend was 340 rubles, and I spent 150 rubles on rent.[9] I lived there until March. The conditions were so bad that at some point I got sick. I could not really study in this basement and usually did my homework at the institute's library. Finally, in March 1955, I got a room in the dorm.

Where did you buy your meals then?

I didn't buy any meals really. I would usually buy a loaf of grayish bread and eat it with sugar and water. For a while, I didn't even know that there was a cafeteria at our institute. After I moved to the dorm, things got better. There were some carts near our dorm, which sold canned cod liver. It was cheap, and we would buy bread and make sandwiches with canned cod liver. At some point, they also started selling canned crab meat. We added crab meat to different holiday salads. All students lived like this.

Rimma Nikolaevna, who or what, you would say, shaped your worldview?

I think it was mostly my family. After all, my grandmother, my mother, and one of my uncles were educated people. My grandmother read a lot and told me about different things. She knew many poems by heart. I read a lot, too. I didn't have much guidance, but I still read extensively. When I think about it now, it appears there was some silent resistance present in my family, even though no one said anything out loud. At the same time, I truly believed what I heard on the radio or read in newspapers. How can this be? What they said was part of Soviet propaganda, but they used all the right words, and we were not really able to critically evaluate these words. We simply believed that what we heard was true. You see, I had a healthy environment at home: no fighting or quarreling really. And I guess I projected this to the rest of society, while growing up. I have never answered questions like this before, but I think my worldview was shaped by my family and my school. The fact that no one in my family lied made me more susceptible to what I heard around me. For example, my grandmother always said: "Put yourself in her or his place and act accordingly." This was a Christian thing to say, even though no one mentioned religion at home. But it shaped my relationships with people. And it was important, because in 90 percent of situations, the way you treat people defines how they treat you. Besides, for a long time, I didn't really have an opportunity to see any discrepancy between official propaganda and everyday reality. I am not sure if I answered your question.

What you are saying is intriguing. Indeed, many words used by official propaganda were positive words.

Yes, they were always positive words. For example, they said they wanted to build a just and fair society. Who would say "no" to that? Or they said that our lives would improve. When I was a child, I believed it. I remember watching *Cossacks of the Kuban*, a propaganda film made in the late 1940s.[10] They screened this film in Quvasoy. The film showed how prosperous Kuban was: how much bread they had and how happy people were there. You should watch it now! But at the time I believed everything I saw in that film. I remember saying: "Well, some people have really good lives." And adults in the room would chuckle at my comment: "They sure do." So they already understood something, and their laughter made me think. But you see back then people didn't really know how to critically evaluate information. Besides, there were not many sources of information. My family would listen to the radio, but we didn't really have money to buy newspapers. In fact, only when I was finishing high school could my family afford a basic radio set.

I have one last question about your life in Central Asia. Even before you came to Moscow for your studies, in March 1953 Stalin died.

Well, yes, I remember this very well. I had a drawing teacher in my high school. I guess he had also been previously repatriated. When Stalin died, indeed, many people gathered outside to listen to loudspeakers. I was at the drawing lesson, and someone from the principal's office came to our classroom to tell us that Stalin had died. Everyone was shocked. But my drawing teacher shrugged his shoulders and said: "Well, he was a senior person, so he died." Some students around me started crying. But I was stunned by my teacher's reaction. Going back to your question about what shaped my worldview, it was things like that. I don't remember if I cried or not, I just don't remember.

And how did your family take the news?

I don't remember specifically. My teacher's words stuck with me, even though his reaction was, in fact, normal. I don't think any member of my family was especially upset. Of course, I didn't get any more pens for my birthday after that (laughs). Times had changed.

Rimma Nikolaevna, at what point did you get interested in science?

I never really thought about science or becoming a scientist when I was in school. In fact, when I was leaving Kyzyl-Kiya to go study in Moscow, I remember telling my chemistry teacher: "Wouldn't it be wonderful if I got to work at a factory one day." But my teacher had worked at chemical factories and she was much less enthusiastic about it. No, at that point I didn't really think about science.

Did you know that a profession like this existed before you came to Moscow? Did you know that one could be a scientific researcher or a professor?

I had heard about professors. At least, I knew such a profession existed. In my family, a distant relative either was a professor or had a similar job somewhere in Moscow or near Moscow. But I didn't really know much about it.

Did you have an opportunity to work as an intern at a research institute later on, when you were a student at the Mendeleev Institute?

No, all our internships were at factories. I studied at the Fuel Department where I was trained as a coker (*koksovik*). Once I had an internship at a factory in Novocherkassk. We had to go from one shop to another, but I don't remember any details. Then I had two more internships in Nizhny Tagil and in Chelyabinsk. When I was already graduating from the institute, I had a choice: to go work at Orsko-Khalilov Combine or at the Irkutsk Branch of the Academy of Sciences. I decided to go to Irkutsk. There were six people in my group who went to Irkutsk. I still didn't know that I would become a scientist at that point, but it was definitely a step in that direction.

So even when you graduated from the Mendeleev Institute, you didn't know you would become a scientist?

I didn't know yet, but I really enjoyed working on my honors thesis. As a coker, I had to work with dispersing and testing resins. Then I had to analyze the results and draw conclusions. So gradually I got more engaged in doing research.

What do you remember about your student life in Moscow? After all, you were a student in Moscow from 1954 to 1959, an exciting time.

First, I actively participated in various sports activities. I was one of the best track runners at my institute. I didn't do sports for sports' sake. I just wanted to have something in common with my peers, to have some common goals. Then in 1956 there was, of course, the Twentieth Party Congress. There were heated debates all over the place. But I still didn't understand much at the time. I think that the Muscovites, who had been affected by this directly, understood much more. They were my peers, yet they had been prepared by their experiences to see the other side of Soviet life. Students from the provinces, however, including myself, didn't know that the other side—repressions, labor camps—existed in the first place. We were astonished, but we didn't understand much. Then in 1957 there was a youth festival in Moscow. It was another big shock.

Did you participate in it?

Well, of course I participated. I represented my institute. The Mendeleev Institute hosted the American and the Polish delegations and organized parties for them. I remember a Polish boy invited me to dance at one party, and it was also something extraordinary for me. There was a very special atmosphere in Moscow during that festival. Many of us didn't own any nice clothes that we could wear during the festival. How did the state solve this problem? They produced an enormous amount of printed cotton skirts and blouses. They were very colorful and pretty, and we all looked dressed up wearing them. They wanted us to look good in front of foreigners. Moscow was flooded by those multicolored skirts. I remember they asked us to vacate our dorm rooms for the festival. They renovated the rooms to accommodate some members of the Russian delegation there. Since I had to vacate my room, I had to go home to my mom's. Still I saw the first parade in Moscow. There was true euphoria in the capital. Afterwards, we felt this way only a couple more times—first, when Gagarin went to space in 1961, and then in 1991 during the putsch, when we defended the White House. Never again did I see people so united. Moscow was exulted. We had never seen foreigners before. Everyone was hugging. It was wonderful.

Would you say it was a moment of unity during the Cold War?

Yes, but no one really thought about the Cold War. It simply didn't matter if there was a Cold War or not. The only thing that mattered was people: specific, individual people. It was truly extraordinary. In general, we didn't focus on ideology much when we were young, either in school or at the institute. They certainly brainwashed us, but we never thought about it on purpose.

Can you tell me about your friends at school and at the institute?

I had a couple of girlfriends at school. When I think about it now, I guess I was a leader of sorts. First, I was in charge of a pioneer section, then a pioneer group. I always tried to organize people around me. It was not that difficult to be a leader in my schools in Quvasoy and Kyzyl-Kiya. I simply tried to imitate what we did in my elementary school in Aleksandrovskoe. I had several friends: Valia Kudashkina, Valia Kokareva, and Galia Yusupova. Galia's father was an Uzbek, and she looked like an Uzbek, even though her sister looked Russian. I am telling you this because when Galia went to apply to an agricultural college in Tashkent, she received the highest grades for all her exams, except the last one. At her last exam, they asked her to bring her birth certificate, which said that she was Russian. And they failed her. Even though she looked Uzbek, they had certain quotas for Russians, and she was not admitted to college because of these considerations. They called it nationality politics.

Were you the only one of your friends who decided to go study in Moscow?

Yes, I was the only one.

What about your friends at the Mendeleev Institute?

I had several friends at the institute. My two best friends were Nila Guliaeva and Valia Galiata. Both were Muscovites. They saw I was all alone in Moscow, and one day Valia invited me to her home. It was the first time I saw a Moscow apartment. It made a tremendous impression on me, even though it was a simple one-room apartment. Then Nila invited me and another girl from Buryatia to a concert at the Moscow Conservatory. After that we visited the Conservatory on a regular basis. Nila had a wonderful mother, who helped us a lot. She often invited us to her apartment and always tried to take good care of us.

As for teachers, we had many talented professors, who also shaped us. We had a great chair of the Fuel Department—Gleb Nikolaevich Makarov. In general, it was a wonderful department, and I was lucky to be part of such a strong group of students. Around that time, the Virgin Lands campaign[11] also began, but unfortunately I didn't go.

Why unfortunately?

Well, of course I wanted to go. In my senior year, we were scheduled to go, but something didn't work out in the end. It would have been an amazing experience: so many different people came together. True, it was hard and challenging, but it was a great experience for young people.

Were you a Komsomol member, by the way?

Yes, I was in the Komsomol. I was in charge of different sports events.

Did you hear Khrushchev's "secret speech" at a Komsomol meeting then?

Yes, they read it out loud. First, they read it at closed party meetings. Then they gathered Komsomol members and read this speech to all of us. But I don't really remember what I thought about it at the time. All I remember is feeling lost, because first they told us one thing, then they started saying the exact opposite.

It must be challenging indeed to readjust so quickly.

Yes, but they are doing it again now. For example, for years they said that Mikhail Luzhkov was a good mayor. No one really paid attention to the horrible Tsereteli sculptures he installed all around Moscow, even though people could see them. But now, all of a sudden, he is a bad person and a bad mayor, and we simply have to accept it without asking any questions.

Are you saying then that young people knew about the repressions before 1956?

No, Khrushchev's speech was a real shock for us. I remember feeling lost: when they removed the foundation, which held everything together, it was not easy to adapt.

Rimma Nikolaevna, you graduated from the Mendeleev Institute in 1959. What happened next?

I went to Irkutsk, to the East Siberian Branch of the Siberian Division of the Soviet Academy of Sciences, to be more precise. There were six people among my cohort who went to work there. Honestly, I was initially a bit apprehensive about going to Irkutsk. It's Siberia after all. But Irkutsk turned out to be a wonderful place. When our train was approaching the railway station, I saw some young people in narrow pants walking up and down the platform. They were the so-called stiliagi.[12] This is how I realized that there was civilization there, too (laughs). When we arrived at the Irkutsk Branch, it turned out that no one expected us or really needed us there. They sent me to work in the Human Resources Department, where I had to go through some paperwork. I spent two weeks doing that. Then I went straight to the director and said: "I came here from over 5,000 km away to work, not to busy myself with paperwork." So they started to ponder what to do with me. Eventually, they offered to have me work at the laboratory of oxidation in Angarsk. So I went to Angarsk.

Angarsk is a city about 50 km away from Irkutsk. You could get there by train. Except that no one really called it a train; it was called "a parcel" (*peredacha*) instead. People called it *peredacha* because there were so many labor camps and prisoners in that area. The laboratory of oxidation, where I was assigned to work, was located on the premises of the Angarsk Chemical Combine. Remarkably, the entire combine was transferred to the Soviet Union from Germany after World War II. In Germany, it produced liquid fuel. They soon found out, however, that the combine could not operate in the Siberian climate: it was simply too cold in Angarsk. So they transformed it into an oil refinery. Later, they would also establish different chemical enterprises on its premises.

I worked in Angarsk for two years. I received a room in the workers' dormitory—a room that I shared with six other people. At that point, I already decided that I wanted to continue my studies and go to graduate school. So I worked full-time at the laboratory and also prepared for my entrance exams. The food supplies in Angarsk were horrible. There was a grocery store in the city, but there was nothing in it except for sole fish and Soviet champagne. Many years have passed since then, but I still remember this image: sole fish and Soviet champagne. There was nothing else available there. I really don't know what we ate. My roommates—all six women—had cans of milk sitting on their window sill. We didn't need a fridge: it was so cold there in winter that windows would freeze inside. Sometimes milk would freeze, too. There was actual ice on top of it. The dorm had one long corridor and a lot of rooms. There was a large stove sitting at the end of the corridor. Everyone took turns cooking meals on the stove. Later I suggested that instead of making individual meals, we make one meal in one large pot. So we organized something like a commune. There were many different people living with me, but we somehow managed to organize a commune.

After a while, my lab received a three-room apartment for its employees. They put a young family in one room, a single man, named Felix, in another one, and another female employee and myself in the third room. This apartment was located on Stalin Street. Of course, we organized a commune there, too (laughs). At some point, we were cooking ramson, because there was nothing else we could buy. Ramson is wild garlic, and you can make salad out of its leaves.

Were there many scientists around you?

Well, there was Natalia Ivanovna Popova, who was head of this lab. She was a graduate of Moscow State University (MGU). Her advisor was a distinguished scientist, Boris Aleksandrovich Kazanskii. There were a lot of good scientists at the Siberian Division of the Academy of Sciences, many MGU graduates among them. As I already mentioned, our lab was not part of the Angarsk Combine, we just rented space there. This was my first job after college. There was a strict working schedule there: we worked from 8 a.m. to 4 p.m. Everyone usually left right at 4 p.m., which later always surprised me: scientific work typically doesn't end at four, it requires longer hours. But I remember in Angarsk if we stayed late, it counted toward overtime. I was one of the few people who came to work there from Moscow. Everyone else graduated from local institutes and universities. For the Siberian Division, our laboratory was quite good. I should give credit here to our director, Natalia Ivanovna. As I mentioned, I worked there for two years.

In Angarsk I soon found out that there were a lot of labor camps and many political prisoners in that region. Felix, my roommate, told me that when Stalin died, there was a large uprising in the camps, directed against the Georgians. Felix and his sister were, in fact, children of "enemies of the people." They lived in Moldavia[13] until their parents were arrested and executed. Felix grew up in an orphanage. Then he graduated from Sverdlovsk University and was assigned a job at my lab in Angarsk. Felix was a wonderful man. He was a chemist, eight years older than us. He lived alone, and I think he actually never got married.

Rimma Nikolaevna, you mentioned earlier that you had not heard about Stalin's repressions before Khrushchev's speech. How did your views change after you arrived in Angarsk and learned about the extensive labor camp system there?

Well, you just know that there are labor camps around you. Sometimes we saw escorted prisoners marching, but it was not really in Angarsk, it was on the outskirts. Angarsk itself looked like a normal, typical Soviet city. There was Stalin Street there. There were modern postwar apartment buildings and stores.

But wasn't it surprising for you to find out that there were so many camps?

No, I guess at that point I already knew about them. I guess by then I was prepared to accept that they existed.

During my second year in Angarsk, I decided to apply for graduate school in Moscow. I had to prepare for my entrance exams. My roommate and I would get up at 6 a.m. to study. We had to be at work by 8 a.m. No matter what happened the previous night, we always got up at 6 a.m. and studied. I had to prepare for my exam in "organic chemistry."

Did you know where you were going to apply for graduate school?

Yes, I was going to apply to the Institute of Organic Chemistry (IOKh). I passed my exams in 1961 and was admitted to work at the laboratory of organic chemistry directed by Liudmila Aleksandrovna Ianovskaia. It was a very good laboratory; many wonderful people worked there. I was twenty-five years old when I started graduate school. Most of my colleagues were a couple years older: anywhere from twenty-eight to thirty-three years old. I began working there in the early 1960s, the pinnacle of the Thaw, and I was fortunate to have been surrounded by outstanding people. One of them, Oleg Serebriakov, went to work abroad, in England, for nine months. When he came back he organized a seminar and was telling us about his trip. We also had private gatherings and sang songs; Bulat Okudzhava's songs were very popular.

What you are describing reminds me of the famous Moscow kompanii.[14]

Well, we had a laboratory, but inside our lab there was a *kompaniia*. Not everyone was admitted to our *kompaniia*. There were two war veterans in our group. I got to know one of them, Sergey Lafit, and it probably helped.

So in 1961 you went back to Moscow to start graduate school. When did you finish it?

In 1964.

Did you defend your dissertation in 1964?

No, I didn't. But I could not stay at IOKh either, and had to find a job. There were different options. Zelenograd was one of them. I recently watched a documentary, called *The Golden Era*, about Mstislav Keldysh.[15] Apparently, under Keldysh, the Academy of Sciences built many scientific centers around Moscow and across the country. Zelenograd was one of them. Another option was Chernogolovka. There were a couple of other academic towns, where I went for an interview. I learned about Chernogolovka from the head of the graduate program at IOKh. She told me I should first meet and talk about this opportunity with Fedor Ivanovich Dubovitskii[16] and Nikolai Markovich Emanuel. There was another scientist at IOKh, Mikhail Lvovich Khidekel, who was also finishing graduate school and was already working in Chernogolovka. I talked to Khidekel first. Then I went to the Institute of Chemical Physics (ICP) —it was just across the street from IOKh—and had an interview with Dubovitskii. After that he invited me to come to Chernogolovka for another interview. I remember the first time I went there. I had to interview with Aleskandr Evgenievich Shilov. He was in charge of the Department of Kinetics and Catalysis at the ICP Branch in Chernogolovka. The department was still in formation, and they were hiring people. I had to wait for Shilov for quite a while. Eventually, I saw a young man with a briefcase. It was Aleksandr Evgenievich Shilov. He invited me to his office and started asking questions. He was not really interested in my most recent research at IOKh, but when I mentioned that I worked on catalysis in Angarsk, he became excited. This is how I began working in Chernogolovka. I was hired to work at Khidekel's laboratory of synthesis.

Did you move to Chernogolovka right away?

No. Even though I was employed at the ICP Branch in Chernogolovka, I was allowed to stay in Moscow for another six months to finish working on my dissertation. I finished it and defended in November 1965. By that point, I had already moved to Chernogolovka.

Did you get an apartment in Chernogolovka as soon as you moved there?

At first, I shared a dorm apartment with four women. It was a two-room apartment: two women lived in one room and three women lived in another room. Since the institute was hiring a lot of new employees, many young people came to Chernogolovka at that time. I was one of the few who had already finished graduate school and defended my dissertation. Most new employees came directly after their undergraduate studies.

When did you get a separate apartment then?

I got a one-room apartment in late 1965, after I defended my dissertation. Then in February 1966 I got married to Rustem Bronislavovich, and we moved into my one-room apartment. We received a three-room apartment in 1967, after our first son was born. The housing situation in Chernogolovka was decent back then. Later I sat on a trade union committee, which distributed housing. We had to distribute housing in prefabricated apartment buildings, and I remember how difficult it was to convince people to move there, because the apartments were not of the best quality. A family of three could easily get a three-room apartment. This didn't last long, though, maybe a couple of years. The housing situation began to deteriorate in the 1970s, when new employees kept coming, but the construction had significantly slowed down.

What was the atmosphere like in Chernogolovka in the mid-1960s?

We also had *kompanii* here. In fact, Rustem Bronislavovich was one of the organizers. We spent a lot of time together, talking, singing. We began meeting less frequently after we got married and had children.

Did you benefit professionally from moving to Chernogolovka?

When I first came to Chernogolovka, I started working on catalysis. But I had small children at the time, and it was challenging. You see, you have to invest a lot into your research to accomplish something. In 1964, an American scientist William Little proposed the possibility of superconductivity in organic polymers. Nikolai Semenov[17] got excited about this theory. There was a meeting of the Academy of Sciences in Chernogolovka in the late 1960s, during which they decided that our institute should research this theory further. It could have led to a number of scientific and technological breakthroughs. At the ICP Branch, they selected three laboratories that would conduct research in this direction: the laboratories of Shchegolev, Khidekel, and Atovmian. To work on this problem, they gathered a vast group of experimental and theoretical physicists. They also needed chemists. I think at some point both Shchegolev and Khidekel hoped that I would become part of their respective teams. But I was getting married, and at some point they asked me directly if I was planning to have children. I said: "Well, of course." I could never separate work from family, and I always knew that I wanted to have children. So they hired another chemist who started working on synthesis of organic superconductors. After that, they gave all support to him. I tried to stay out of their way, but I continued to research the same problem.

So did they stop supporting your research because you had family and children?

Well, not exactly. Of course, if I had had outstanding scientific results, they would have paid more attention to my research. But it took a while until I got my first significant results. At first, I worked on this mostly by myself, synthesizing different substances. It was hard and dangerous work. I would usually come home, take off my clothes and hang them outside on the balcony. It was impossible to get the terrible chemical odor out of my hair. I wore a mask at work to protect myself, sometimes even a respirator.

There was tremendous competition in this field, both at our institute and abroad. Of course, Western scientists had much better conditions. They simply would not have survived as scientists if they had to work in our conditions. For example, I remember once we went to the Institute of Physical Problems. There was a seminar there, and the presenter was telling us about personal computers (PCs). They had been using PCs for quite a while. By contrast, it was the first time I had even heard about them. It was challenging to find even a typewriter with a Latin alphabet at our institute. When I had to type up the list of foreign literature I used in my research, I had to go to the First Department[18] and ask them if I could borrow a typewriter. They kept taps on all typewriters with a Latin alphabet. We wasted a lot of time on things like that.

Did you feel isolated from the international scientific community?

Yes, but this isolation came to an end around the mid-1980s. Shortly after the launch of perestroika, we participated in the first Soviet international conference in our field. It took place in Tashkent, and they invited some American scientists to take part in it. There were many other Western scientists there. It was the first time we presented a poster with our results. And at some point our poster got attention from a famous French chemist.

Did you have any contacts with scientists abroad before perestroika?

No. In our field Igor Shchegolev, the head of one of the labs at our institute, could travel abroad. Sometimes, Rimma Shibaeva went with him. They usually traveled to conferences. Everyone else who worked in our field didn't go abroad.

Did you have access to Western scientific journals before perestroika? Could you follow how your field developed in the West?

It was extremely complicated. We didn't have access to Western journals. Instead, we received Soviet reference periodicals, which had abstracts of different articles. But these reference periodicals always came late. When I think about it now, I cannot help but wonder: how could we work in such conditions? It was even more difficult to get reagents for chemical reactions. We had to run around, searching for them. It got even worse in the late 1980s, when everything started falling apart. It became impossible to buy anything. When I started traveling abroad, already after 1991, I realized that, in comparison to Western European countries or Japan, we worked in horrible conditions.

Rimma Nikolaevna, how challenging was it for you to combine having a family and small children, on the one hand, and a scientific career, on the other?

It was not easy at all. But I needed both, so I had to find time for both. Once I remember my younger son got sick, he had a high fever. But Rustem Bronislalovich and I had to leave for a conference in Tbilisi that day. I didn't know what to do. On the one hand, I applied for this conference and didn't want to let down people who were relying on me. Additionally, I needed to present my research. On the other hand, my son was sick and he needed me. My friends helped me out a lot then: we left and they took care of our son. But as soon as we got to Tbilisi, we started to look for a telephone so that we could call and see how he was doing.

And what was the attitude toward female scientists in the profession at the time?

Well, the scientific profession was dominated by men, so women had to constantly prove their worth. Sometimes women had to work harder than men to convince their colleagues that they were worth something. For example, I only defended my doctor of sciences dissertation when I had excellent results and no one could question my work.[19] I believe this partially comes from the different nature of men and women. Women are more thorough and methodical. They are more tolerant and more compliant. Men are more interested in getting results and being rewarded. I am not sure if it is good to be so compliant. But I do think a lot of scientific experiments could not be carried out without women.

When did you defend your doctor of sciences dissertation?

In 1991, when my children were already grown-ups.

And when did you establish your own laboratory?

Around the same time. For a long time, I worked in the laboratory of Mikhail Khidekel. It was a large lab: it had more than forty people. But when things started falling apart in the early 1990s, Khidekel decided to transfer his lab from our institute to Moscow. I said I would not move to Moscow, and my group and I stayed in Chernogolovka. Later Khidekel emigrated to Israel, and some people from his lab came back to work with me. I had eighteen people working under my supervision. And at some point I said: "Enough, I need to have my own lab." Once I got my own lab, I began recruiting graduate students. I admitted two young women who had just graduated from the Mendeleev Institute and a young man from Saint Petersburg. This young man still works at our institute.

Rimma Nikolaevna, you lived in a closed scientific town for several decades. How well supplied was Chernogolovka in the Soviet era?

It was not supplied that well. I came here in late 1965, and even though there was already a large grocery store in Chernogolovka, there were few items available in it. When I lived alone, this didn't bother me. But after I got married and had children it became a real problem.

Didn't Chernogolovka have special provisions?

No, not really. Perhaps there were special provisions in other closed towns, but not here. At some point, in the 1980s, the employees of the ICP Branch organized parceled food deliveries. I remember we each got several bottles of vodka as part of these parcels. We didn't drink it, but used it as "currency," so to speak. One could give a bottle of vodka to an electrician, for example, to thank him for his work. It was also difficult to buy any fruit or vegetables here.

What about furniture? How soon could you buy furniture for your new apartment?

When I got married in 1966, there was a lot of furniture from socialist countries available for sale in Moscow. But you could not just go to a store and buy it. There were waitlists, and we had to sign up. My mother gave us some money for our wedding, and Rustem Bronislavovich and I used it to buy Bulgarian furniture. We still have it, by the way. We also got kitchen cabins as our wedding gift. Our salaries, however, were never enough to buy something extra like furniture. Overall, produce and other supplies in Chernogolovka were much worse than in Moscow. One had to constantly search for food or clothes. Sometimes I would see people in line fighting with each other in order to buy something. It was horrible. I never thought it was worth it and never participated in it, maybe because I didn't have time, or money for that matter. My goal was to provide my family with some basics.

Did you participate in organizing cultural life in Chernogolovka?

No. Rustem Bronislavovich did, but I had other priorities: my work and my family. I simply didn't have time for other things. But I was a member of the House of Scientists[20] and paid my regular membership fees. We had wonderful cultural events there, and I attended many of them. For example, there was a cinema club in Chernogolovka. They invited many famous artists, like Sviatoslav Richter, to perform in our town. We also were members of the Moscow Conservatory and regularly traveled to various theaters in Moscow.

Would you call Chernogolovka "an oasis of intellectual freedom"?

I think I would. You see, many smart people came to live and work here, and they formed the foundation of the local community. If you compare Chernogolovka with an average Soviet town, of course, we had more intellectual freedom here, simply because people who lived here were critically-minded, intelligent people. In general, I am grateful that I could live my life in such a wonderful place: I could do what I loved most, science, and it was so close to Moscow.

But Chernogolovka never really had any informal clubs or organizations, like Akademgorodok did.

I cannot compare Chernogolovka with other scientific towns, since I don't know them well. What I do know, however, is that Fedor Ivanovich [Dubovitskii] could not change the Soviet system all by himself. If he had openly demanded more intellectual or political freedoms, he would have had to go against the system. There was a lot of pressure on him from above, and he had to maneuver to protect his scientists and his town. He was a very humane person, but he alone could not change the system. Only in 1991, when millions of people went outside in the streets, could something change. There was some hope then. . . . You see, we didn't really think about any of this in the 1960s or 1970s. Scientific work usually required total dedication.

Rimma Nikolaevna, where did you get information from in the 1960s and 1970s? What newspapers did you read? What radio stations did you listen to?

We got information mostly from our friends. One of them was Kronid Liubarskii.[21] Liubarskii also gave us samizdat literature. I remember reading Solzhenitsyn's *Cancer Ward* and feeding my elder son. We also listened to foreign radio broadcasts, such as the Voice of America.[22]

Were you a member of the Communist Party?

No, I was not. I was, however, an active member of Komsomol. When I came to work in Chernogolovka, they tried to convince me to become a secretary of the Komsomol organization at our institute. But I refused, because I wanted to focus on my research. When I turned twenty-eight years old, I received a recommendation to apply for party membership. Martemianov, Liubovskii, Petinov, and I—we all got recommendations and applied. We had to wait, but when it was finally time to go to Moscow for an interview, I was already six months pregnant, and I simply didn't go. Later, they offered me to join the party several times, but it never worked out. Thank God!

Why did you apply for party membership in the first place?

Our main goal was to try to change the party from within. But I am really glad now that it didn't work out for me, and I didn't join the party. Taking care of my family was more important. Besides, it soon became obvious that it was hardly possible to change anything. One could invest a lot of time and efforts into trying to change the system without any noticeable results.

Did you have any specific ideas as to what had to be changed?

I don't think I did. Perhaps my male colleagues had some ideas. Really, I had to deal with the party organization directly only once, in 1986, when I needed to get their approval for my research trip to Poland. I got to know a professor in Poland, through publications, and he invited me to come and work at his institute. I had to apply through the Academy of Sciences. I had to prepare a long list of documents, including my application, autobiography, characteristic, and information about my relatives. I had to submit six copies of everything. Then I had to pass a special commission in Chernogolovka, which then sent my documents to Moscow. If your documents were approved, you were invited to a party organization committee, which asked you different questions. They could ask if you participated in the public life of your institute or who the head of some socialist country was. I remember they asked me questions about some officials of the Polish Workers' Party. I didn't know the answer, and a young woman on this committee began to scold me. It lasted for quite a while. It was really humiliating, as it had nothing to do with my research. At the end of the day, they approved my candidacy, but it was an extremely unpleasant experience. Overall, few people were allowed to travel abroad at the time—mostly those who had a connection at the Foreign Department of the Academy of Sciences. This didn't change until the 1990s.

How important were these contacts with the international scientific community?

They were extremely important. We traveled mostly to conferences inside the Soviet Union; some people traveled to international conferences. But if you didn't go at all, you were eventually left to your own resources.

Have you ever considered emigrating to another country?

No, I didn't. When we were still students, some girls at the Mendeleev Institute married foreigners from socialist countries. But I never even considered such a possibility. I guess I was raised in a certain way. Later, after perestroika and the Soviet collapse, there was a lot of interest in Russian scientists abroad. At that time, we already had good scientific results, and we could have left, if we wanted to. But I never considered it seriously, even though life here became extremely difficult. I guess the fact that I don't know English well was an obstacle. Then my family was here: I could not leave without Rustem Bronislavovich. Although at one point, both he and I received an invitation to come and work in Slovenia for six months. It was in the early 1990s. It was an extremely challenging time for our country. Ironically, it was the most productive period in our scientific careers. We had a lot of contacts with scientists abroad, who knew our work. In the 1990s, we traveled to Sweden, Japan, the United States, and France. From a professional standpoint, it was an amazing time.

What did you know about Western culture and the West in the 1960s and 1970s?

We knew very little. I knew very little. Some people say that they were influenced by Western music, for example, *The Beatles*. I was not one of these people.

Did you feel the impact of the Cold War in your everyday life?

You see, Communist ideology was ingrained in our minds, but so was fear of war. Since I lived through the war, and since I knew what it was like from my personal experience, I always had this fear. Our ideology focused on one key goal: to avoid war in the future. Since I survived wartime bombings, to this day the noise of military planes makes me anxious. I guess all children of war have this fear deep inside them. I remember the Caribbean crisis, for example, and how terrifying it was. But we lived like this. Some tension was always present.

Did you feel the effects of détente in the 1970s?

Well, we welcomed any relaxation of tensions, but we didn't think about it specifically.

Did you perceive the West, and the United States in particular, as hostile countries?

I don't think so. These tensions concerned politics and had little to do with ordinary people in these countries. But I can only speak for myself. I remember when Aleksandr Shilov returned from one of his trips abroad, from the United States in fact, he organized a seminar. After the seminar, he was telling us about his impressions. He said he saw caricatures of US presidents. We were all shocked. How could they do this? No one could have imagined having a Brezhnev caricature displayed in public view.

Rimma Nikolaevna, let's now talk about dissidents. How did you meet Kronid Liubarskii?

I met Kronid through Rustem Bronislavovich. I knew that my husband was working with him on translating a book. They would usually meet at Kronid's apartment. One day I came with my husband. I remember Kronid was making toasted bread and serving it to us, since he had nothing else in the house. I cannot say that he made any particular impression on me. I had two little children and my work, and that was more than enough. I didn't communicate much with Kronid when he lived in Chernogolovka. I remember the New

Year just before he was arrested. On New Year's Eve he came to our place with a big bag full of presents. We put on some masks and went from one apartment to another like that. It was raining outside, and we had umbrellas with us. It was a lot of fun. Shortly after that Kronid was arrested and I barely communicated with him after that.[23]

Did Rustem Bronislavovich tell you that Kronid gave him samizdat to read?

Well, of course, he did. I also read it, but I didn't know Kronid so well before his arrest. I knew there was a search of his apartment, and then another one. I knew that his wife, Galia, visited him in prison in Vladimir.

Were you shocked when you found out about his arrest?

Of course I was, even though I knew he was a dissident and that it was dangerous. I also knew that we put ourselves in danger, too, by simply reading this literature. It was clear to us that we were reading banned literature. After Kronid's arrest, we began to hide it better. Besides, we knew about the trials of other dissidents. We lived in Chernoglovka, where people discussed these kinds of things. Of course, we knew that Kronid was distributing literature that was officially banned, but we read it anyways.

Are you saying that local scientists knew about the trial of Sinyavsky and Daniel[24] *and other dissident trials?*

Yes, of course they did. Scientists talked about it. We recognized that dissidents were brave people, and we sympathized with them. We also recognized that they were doing the right thing, speaking up against the established order.

Does it mean that you had doubts about the established order as early as the 1960s and 1970s?

How could you not have doubts, when they tell you that you have access to everything, but you cannot even buy some basic products in the stores?

Do you remember if you had these doubts when you were still a student at the Mendeleev Institute?

No, I didn't. I guess I lived in a different kind of environment. I think my eyes began to open later, and it happened gradually.

Was there any one event that accelerated this process?

Yes, there was—the arrest of Kronid. First, he was arrested. Then the KGB started bringing people who had known Liubarskii in for interrogation. I was preparing myself. It was terrifying, really. I had two small children. The KGB's goal was to prove that Kronid had distributed samizdat. The interrogations took a long time, and I was afraid that I would not be able to consistently deny that Kronid had given us samizdat. At the end of the day, they didn't summon me, although my husband was summoned, and I was worried about him.

Could they arrest you in 1972 for reading samizdat?

Of course they could.

For reading, not distributing samizdat?

You know the distinction between reading and distributing was always blurred. Kronid reprinted samizdat on his personal typewriter. He could only produce five copies at once. So you could not distribute a lot like that, but that was enough for the KGB to put him in jail.

What did you think about your husband's friendship with Kronid? Did you realize it could be dangerous?

At that point, we already knew that Kronid was standing up for the right thing. You see, we always listened to the Voice of America. Then the Soviet invasion of Czechoslovakia took place. It was awful. My husband even got in minor trouble at work for discussing the events in Czechoslovakia in his lab. But this was the way we lived back then. We knew that we didn't violate the law, yet we also knew that reading samizdat could get us in trouble.

Do you think that Kronid's dissident activity in any way threatened the Soviet way of life or Soviet power? If so, I have an even more provocative question for you: was the Soviet state right to put him on trial?

The trial of Kronid was wrong and unjust. It was obvious to us. But the Soviet regime believed that his activity threatened the state. "From a spark a fire will flare up" (*Iz iskry razgoritsia plamia*).[25] I guess they were afraid that if dissidents enlightened the people, this could be dangerous for Soviet power.

Did scientists discuss the Liubarskii Affair at the institute?

They didn't discuss it at the institute, but everyone knew about it. We were young and socialized a lot. Of course, most of our conversations took place in the kitchen. People in my circle thought that Kronid's arrest was terrible, unjust, and scary.

Did you continue to communicate with Liubarskii's family after his arrest?

Yes, we did. His family lived in Chernogolovka until Kronid was released from the Gulag.[26] But we were not very close with them at the time. We became close friends after Kronid and his family returned from emigration in the 1990s. It was an exciting time, and we always had guests arguing about different topics in our kitchen. Kronid was one of our guests. He would usually come with his wife, Galina. Once I asked him, during one of these conversations: "Kronid, what does it mean to be a dissident?" I began to think that it was really a way of life. And since people are different and they have different ways to live, I thought one should not condemn a person for not being a dissident. I asked him if I was right, if it was really a way of life, and Kronid said "yes." I felt a connection forming and I hoped I could learn more from him, but soon after that Kronid died. Trying to support his wife Galina, we became close with his family.

Why did I ask Kronid this question? You see, the dissident circle condemned other people for being conformist, for not being dissidents. And I thought they should not have done it, because not everyone could become a dissident. Not all people could feel and think as they did. Theirs was a closed world, although different people ended up as dissidents for different reasons. But they all made a choice to become dissidents. It was their choice, and I thought they were wrong when they tried to impose their choice on other people.

Rimma Nikolaevna, did you believe, at any point, that it was possible to build a Communist society in the Soviet Union?

Well, I remember they promised we would all live under Communism in twenty years. When I was little, I would ask my mom: "What does it even mean 'From each according to his ability, to each according to his needs?'" I had a hard time understanding this slogan. Since we were extremely poor, I could not understand how this could be accomplished. I would ask my mom: "Does it mean that people won't need to work and can still have everything?" I was little, of course. When I grew up, these questions no longer interested me.

But did you actually think, at some point, that you lived in a socialist society?

I don't think I did. I guess in theory the Communist ideology was not that bad, but as you grow older, you gradually understand that it's a utopian ideology. Khrushchev said that by 1980 we would all live under Communism, that everyone would get a separate apartment. But this didn't happen. The year 1980 arrived, but people were still waiting for their apartments.

Yet I think that what we have now—call it capitalism or something else—is also bad. I think in the 1990s everyone had his or her own vision of our country's future. But these visions had little to do with reality. No one really knew what would happen when we removed old Communist officials from power. I think this is the tragedy of every revolution: it is not enough to go into the streets and remove the current leadership from power. It is simply not enough. Everyone wanted to have some pure, honest people in charge, whose actions would support their words and promises. But this didn't happen.

Would you say that the scientific intelligentsia was pro-Western or that it hoped for more elements of democracy or capitalism in the USSR?

I can only speak for myself. I think that we simply didn't know what life in the West looked like. We were naive and idealistic. We had some ideas, but they were not based on our personal experiences. Nowadays, when we got to know life in the West much better, I am not convinced that this is an ideal way to organize society or that we need to aspire for that in our own country. Money rules the world; the rest is just words that people use to disguise this reality.

What about the Soviet era? How was it different?

Well, the economy is of primary importance in every society. The economy had to be a priority in a socialist society, too. The problem was that we had central planning, and it was awful. It was all about fulfilling the plan and reporting your results to a central agency. No one cared about actual profits. Institutions on the ground had to ship off their products to the center, where it got redistributed. But this socialist economy didn't make any sense, because any housewife will tell you that she would rather buy what she needs at a local grocery store or market, rather than go somewhere else to look for it.

And what did you think about Gorbachev's coming to power in 1985?

I was born in Stavropol'skii krai, and Gorbachev worked there for a while. So when he came to power, I viewed it quite positively. He was a relatively young man, only several years older than me. Gorbachev came to power because the older generation was dying out. So his reforms were part of this generational break, and we expected a lot from them.

You see, I feel bad about the collapse of the Soviet Union. But nowadays when people start arguing that Yeltsin and Gorbachev were to blame and that they could have prevented the collapse and saved the Union, I disagree with them. I don't think Yeltsin and Gorbachev were to blame for the collapse. Perhaps they could have tried to save the Union. But there were terrible regional wars all across the USSR. The real problem, I think, was that there were local "tsars" at the head of every Soviet republic, and they wanted to have more power. No one really thought about the consequences of the Soviet collapse either for our country or for the world. It was natural for the Baltic States to leave the Union, but all other republics also wanted to become independent—not so much the ordinary people, but people in power.

I was always a strong supporter of Gaidar, Chubais, and their team.[27] They were relatively young when they came to power, in their mid-late thirties. But now I am starting to think that they tried to imitate the West too much. Perhaps Russia needed something else. The situation back then was extremely complicated.

Do you think Gorbachev was right to have launched glasnost[28] *and the political reforms?*

I think he was right. Introducing glasnost was the right thing to do. Of course, the answer to this question depends on whom you ask.

Some of your colleagues argued that Gorbachev should have started with economic reforms, keeping tighter control over the political sphere.

I don't think he could have kept tighter control over the political sphere, but I never thought Gorbachev was a particularly wise man. The fact that his wife always interfered in politics didn't do much good either.

Rimma Nikolaevna, you mentioned earlier that you were actually present at the August 1991 events in Moscow.[29] *What made you go there?*

The August 1991 events were, in fact, quite terrifying. Rustem Bronislavovich might have told you that we were at the *dacha*[30] when we found out about the putsch. I remember he woke me up at dawn, saying: "You will not believe what is happening in Moscow." We didn't really know what was going on. Our dacha is 25 km away, close to Dubrovo. If our Soviet past was any kind of indicator, we had good reason to believe that, since we were Yeltsin's supporters, they would start persecuting us. At that point, Rustem Bronislavovich was a member of the town council in Chernogolovka. It was an extremely turbulent time. People got together and discussed how to organize the town council, how to elect institute directors and other things. The goal was to elect honest, energetic, and nonpartisan people into the town council. They accomplished this goal. The only problem was that they were mostly scientists, whose main priority was science. They were not professional economists or politicians. Later, scientists would be forced out of the town council, and more profit-minded people would take their place. But at the time of the putsch, Rustem Bronislavovich was still an active participant in the town council.

When we learned about the putsch, our first thought was that the military had probably blocked off the road from Dubrovo to Chernogolovka. But our children were in Moscow, and we decided that we had to go back right away. Before that we heard Yuri Afanasiev's talk on the radio,[31] in which he confirmed that this was indeed a putsch, but said that we would win. It made us feel a bit better. In his speech, Afanasiev asked people to come to the White House.[32] So we decided that, first of all, we had to go back to Chernogolovka and then probably to the White House. To our surprise, we found out that the road to Chernogolovka was clear. Rustem Bronislavovich immediately went to the town council. It was August 19. The participants of the town council didn't really know how to proceed. Some people thought they should go to the White House and defend Gorbachev and his reforms. My husband was one of these people. Others disagreed. No one understood what was happening. People were too afraid to act, since the outcome of the putsch was still far from clear.

On the night of August 20–21, my friend Galia Shchegoleva called me and said that they were going to the White House and that there was space for one more person in the car. This is when I decided to go. We left late in the evening to get there before the curfew began. We parked our car somewhere and walked to the White House. It was an incredible night. It is hard to describe it now. There were tons of people everywhere. People were standing around bonfires; they were crowding on the stairs of the White House. There was something surreal about it, even absurd. I remember announcements from loudspeakers: "If you have sandwiches or gauze, please bring them to us." People were everywhere, walking up and down the streets near the White House. The four of us—Galia, her husband, their friend, and I—also started walking. I was, in fact, looking for my children. I knew they were there, and

I wanted to find them. I saw young people holding hands, making a human chain. I walked very close to them, looking into their eyes. I will never forget those eyes! God forbid, if they had given orders to send one or two tanks there, if they had given orders to shoot at people, they would have created such a massacre. These people were completely unarmed, but if they started shooting, no one would have left or moved.

There were different people there that night: social stratification had already begun by then. There were some businessmen among them, there were students, but they all formed one united front. Igor Shchegolev suggested that we go to the tunnel. We saw a burning bus and people trying to put the fire out. There were lots of people in the tunnel, too. No one knew what was happening. Rumors began to spread that they were sending helicopters and tanks and that they were going to bomb the White House. There was incredible tension in the air. It was terrifying. We decided to walk to the bridge. There were human chains there, too. People were preparing for an attack on the defenders of the White House. There was a lot of confusion. And fear. We stayed there until dawn. At dawn it became clear that there would not be a full-fledged attack on the White House, and we decided to drive back home. When we arrived in Chernogolovka, there was a mass rally there, and they asked me to talk about my experiences at the White House.

Did you find your children near the White House?

No, it was impossible, but this is exactly why I went there. It was easier for me to be there than to stay at home. Our children were there—Pavel, Maksim, with their friends. It was easier for me to be there and see everything with my own eyes than to wait here.

Why do you think all these people went to the White House in the first place?

They went there to defend their truth, their faith, and their individuality. Everyone had his or her own version of truths and their own ideas about how the reforms should progress, and they went to defend them.

*Rimma Nikolaevna, do you consider yourself to be a part of the so-called Thaw generation, or "people of the sixties" (*shestidesiatniki*) generation?*

Well, of course I do. I think "people of the sixties" sounds a bit pretentious. It usually refers to artists. Yet we definitely belonged to the generation that had been shaped by the Twentieth Party Congress.

NOTES

1. Aleksandrovskoe was occupied by the Nazis on August 7, 1942, and liberated by the Red Army on January 14, 1943.

2. The drought caused a bad harvest failure, which was exacerbated by the mandatory state procurements of grain and led to the 1946–1947 Soviet famine.

3. The All-Union Pioneer Organization, established in 1922, was a Soviet youth organization for children aged nine to fourteen years old. Its main purpose was to indoctrinate Soviet youth in Communist ideology and to prepare young people to apply for membership in the Communist Youth League (*Komsomol*), which was the final step before joining the Communist Party.

4. Quvasoy was a city in the Fergana region in the Uzbek Soviet Socialist Republic.

5. Cracovienne, or Krakowiak, is a traditional Polish dance from the region of Krakow.

6. A gold medal guaranteed a student admission to most Soviet colleges, without passing any entrance exams.

7. Present-day Bishkek, the capital of the Kyrgyz Republic, or Kyrgyzstan.

8. "Third shelf" was not technically a passenger seat; its main purpose was to store luggage.

9. Liubovskaia's stipend was equivalent to 34 rubles after the monetary reform of 1961.

10. The film, glorifying the life on collective farms in the Soviet Kuban region, was officially released in 1950.

11. The Virgin Lands Campaign, launched by Nikita Khrushchev in 1954, aimed to solve the agricultural crisis in the Soviet Union by bringing millions of hectares of previously unused land, mostly in Northern Kazakhstan and Western Siberia, under cultivation. The campaign relied heavily on Soviet youth: in 1954, for example, about 300,000 Komsomol members volunteered to participate in the Virgin Lands.

12. *Stiliagi* (lit. "style apers"), or hipsters, were part of the Soviet youth subculture in the late 1940s and 1950s. They dressed in Western-style clothes and listened to American music, defying the Soviet behavioral canon. For more information on *stiliagi*, see Vladislav Zubok, *Zhivago's Children: The Last Russian Intelligentsia* (Cambridge, MA: The Belknap Press of Harvard University Press, 2009), pp. 40–44.

13. The Moldavian Soviet Socialist Republic was established in 1940, shortly after the Soviet annexation of parts of Bessarabia from Romania. After the dissolution of the Soviet Union, it was renamed the Republic of Moldova.

14. *Kompanii* were informal groups, usually circles of friends, which became widespread during the Thaw. They consisted mostly of young educated people, who got together to talk about ideas and the fine arts.

15. Mstislav Keldysh, a prominent Soviet mathematician, was president of the Soviet Academy of Sciences from 1961 to 1975.

16. Fedor Dubovitskii was one of the founders of Chernogolovka (together with Nikolai Semenov). In 1962, he was appointed "the Authorized Representative of the Presidium of the Soviet Academy of Sciences in Chernogolovka."

17. Nikolai Semenov was the founder and first director of the Institute of Chemical Physics; he was also one of the founders of the scientific center in Chernogolovka.

18. The First Department, part of the KGB, existed at most Soviet institutes and enterprises. It controlled access to information, which was considered to be secret, as well as foreign travels and publications.

19. Doctor of sciences (*doktor nauk*) is the second, highest doctorate degree in Russia and the former Soviet Union. It can be earned after a candidate of sciences (*kandidat nauk*) degree, which is roughly equivalent to a PhD.

20. The House of Scientists was a cultural center. It hosted concerts, lecture series, and performances.

21. Kronid Arkadievich Liubarskii was a dissident scientist who lived in Chernogolovka from 1967 to 1972. During this time, he befriended Rustem Liubovskii, Rimma Liubovskaia's husband, who borrowed samizdat from the dissident.

22. The Voice of America was part of the US propaganda campaign against the Soviet Union during the Cold War. It began its first Russian-language radio broadcasts in 1947.

23. Kronid Liubarskii was arrested in January 1972 as part of the KGB campaign to stop the production of the *Chronicle of Current Events*, notoriously known as Case 24. He was accused of "spreading anti-Soviet slanderous fabrications with the purpose of undermining and weakening Soviet power" (Article 70 of the Soviet Penal Code), put on trial in late October 1972, found guilty and sentenced to five years in a labor camp.

24. Andrei Sinyavsky and Yuli Daniel were two Soviet writers who had used pseudonyms to publish satirical short stories in the West. They were charged with spreading "the anti-Soviet propaganda" and put on trial in 1966. Sinyavsky was sentenced to seven years in a labor camp; Daniel was sentenced to five.

25. This line from Aleksandr Odoevskii's poem became the epigraph to *Iskra* (*Spark*), which was an underground newspaper founded by Vladimir Lenin in 1900.

26. Liubarskii was released in 1977. He and his family were forced to emigrate shortly after that.

27. Yegor Gaidar was Russia's Prime Minister and the architect of the controversial "shock therapy" reforms in 1992, which resulted in the privatization of Russia's state-owned enterprises. While the reforms facilitated Russia's transition from the Soviet-style command economy to a market economy, they were incomplete and led to massive hyperinflation and the significant deterioration in living standards in post-Soviet Russia. Anatoly Chubais was a member of Boris Yeltsin's administration in the early 1990s and another avid proponent of the privatization campaign.

28. The policy of glasnost, introduced by Gorbachev in 1986, called for more openness in discussing public affairs, as well as the Soviet past.

29. The August 1991 Putsch, or the August Coup, was an attempt by the State Committee on the State of Emergency (GKChP), to remove Mikhail Gorbachev from power and to reverse his reforms. The coup failed within two days.

30. A country house in Russia, usually used as a second home.

31. In 1989, Chernogolovka played a crucial role in getting a prominent historian Yuri Afanasiev elected to the newly convened national assembly, the Congress of People's Deputies.

32. The White House was the seat of the Russian government. During the August 1991 Coup, Boris Yeltsin went to the White House and denounced the State Committee on the State of Emergency as unconstitutional.

Chapter 2

"I Was Only Ten Years Old When My Father Was Arrested"

Vladimir Karlovich Enman

One of the first residents of Chernogolovka, Vladimir Karlovich Enman (b. 1928) came to the town in March 1959 to accept an engineering position at the Branch of the Institute of Chemical Physics, and has lived there ever since. In his early eighties, Enman looked surprisingly youthful. Slender, energetic, and quick in his movements, he arrived early to our first meeting on August 5, 2009. After my brief explanation of the interview framework, Vladimir Karlovich started talking. His speech was hasty, even hectic at times, as if he had waited his entire life to tell his story. When the narrative became confusing, I tried to get it back on track with questions. For the most part, though, I just listened. I conducted two follow-up interviews with Enman on October 3, 2010 and March 3, 2012, to clarify some of the themes that emerged out of the first interview.

The arrest of Vladimir Karlovich's father in June 1938 was the major event that shaped his identity and his worldview.[1] *Karl Karlovich Enman built a successful career in the Soviet military, which he had joined during the Russian Civil War. After spending nearly a decade in Ukraine, he was transferred to Moscow in the early 1930s, where he taught courses to commanding officers of the Mechanized and Motorized Troops of the Red Army. In 1935, a thirty-nine-year-old Karl Karlovich was promoted to head the Department of Firing Instruction at the Automotive and Armor Directorate of the Red Army. His good luck, however, didn't last long. In June 1938, Karl Enman was arrested during the mass purges of the Soviet military. He was charged with sabotage and espionage and executed on September 27, 1938.*[2]

The execution took place at Kommunarka, a shooting ground right outside of Moscow. Yet, Enman's family knew nothing about it. Shortly after the arrest, the entire family was evicted from their luxury two-room apartment in Lefortovo and squeezed into one room of a communal apartment located in the basement. All of their belongings were confiscated. Vladimir Karlovich was only ten years old when this happened. Since the arrest and the search took place at night, he slept right through them. After his father was "taken," Enman's mother was too afraid and shaken to tell her children about what had happened. Vladimir Karlovich learned about his father's arrest by accident. One day he was playing outside with some local boys, and one of them called him "a son of the enemy of the people."

The trauma inflicted on Enman by his father's arrest was so extensive that it arguably overshadowed the suffering and starvation that he experienced during the Great Patriotic War. When time came to apply for college, he hesitated to submit his documents to any schools in Moscow, terrified that the stigma of having a repressed father would land him in jail, too. Committed to attending college no matter what, Vladimir Karlovich eventually decided to leave Moscow and traveled to the Latvian Soviet Socialist Republic, where he successfully applied to Riga River College. He got married in 1955. But even then, Enman concealed his family's tragic background from his wife. He only told her the truth after his father was officially rehabilitated in August 1956.

Like millions of Soviet youth who lived through the mass repressions and the horrors of World War II, Vladimir Karlovich was convinced that getting a good education was his only chance to succeed in Soviet society. He was a passionate and hard-working student, which helped him do well in school. In 1950, he graduated from Riga River College with distinction, receiving a qualification of a marine mechanical engineer. From 1950 to 1955, he went on to study at the Leningrad Institute of Marine Transport Engineers. Despite Enman's excellent academic record, in 1955, he didn't obtain the security clearance required to work at Soviet research facilities, and had to take a job at the Marine Engineering Plant in Kaliningrad.

It was the death of Joseph Stalin in March 1953, and the ideological relaxation that followed, which allowed Vladimir Karlovich to overcome the stigma of his family's past. In August 1956, the rehabilitation papers arrived, stating that his father was innocent and that he was cleared of all charges. To this day, Enman believes that his professional success would not have been possible without the rehabilitation of his father. Liberated from the stigma he experienced in his youth, in 1956 he got a job at the Kaliningrad Branch of the Laboratory of Engines of the Soviet Academy of Sciences. The laboratory was directed by Boris Sergeevich Stechkin, a prominent Soviet scientist who made significant contributions to the construction of Soviet aircraft engines. Vladimir Karlovich worked as a senior engineer at the Kaliningrad Branch from 1956 to 1959, when he decided to move closer to Moscow. In March 1959, he brought his family to Chernogolovka, where he got a job as an engineer at the Branch of the Institute of Chemical Physics.

Enman took a lot of pride in his work and his professional accomplishments. In 1961, he was appointed the main engineer of the ICP Branch. He organized and managed the institute's machine workshop and the Construction Bureau, overseeing nearly three hundred people. His story, like the other oral histories in this volume, sheds light on the lives of the professional elite in the late USSR. In the 1960s and 1970s, Enman was able to provide a relatively high standard of living for his family. He received a separate two-room apartment shortly after his family moved to Chernogolovka. In 1963, Enman's family moved into a spacious three-room apartment with high ceilings, wooden floors, and its own kitchen and bathroom, which was pure luxury even after Khrushchev launched the massive housing construction program in the mid-1950s. Vladimir Karlovich felt incredibly lucky to be a part of the scientific intelligentsia, which enjoyed such outstanding privileges.

Yet, Enman never forgot the tiny dark room in the basement, where his family was confined after his father's arrest. Even though he was grateful to the post-Stalin state for the professional opportunities he received in the 1960s and 1970s, he held it accountable for his father's unjust execution and his family's misery while he was growing up. For a long time, Enman's resentment precluded him from joining the ranks of the Communist Party. Even after he finally joined in 1968, he could never fully identify with the party's official ideology and was skeptical of its utopian promises. At the same time, like many children of the repressed, Vladimir Karlovich was cautious to hide his skepticism. When Stalin died, for example, he joined his classmates on a train trip to Moscow to say goodbye to the deceased dictator. Even though he sympathized with the dissident movement, he didn't read samizdat or listen to foreign radio broadcasts. He kept his doubts to himself until Gorbachev's perestroika, when information about the scale of Stalin's repressions and the Soviet regime's crimes became widely available.

In the early 1990s, Vladimir Karlovich found courage to go to Lubianka, the notorious prison and the headquarters of the KGB in Moscow. He requested to see his father's dossier. When I interviewed Enman twenty years later, I was struck by his determination to remember his family's history, as well as his lack of nostalgia for the Soviet past, so typical for many members of his generation.

My father, Karl Karlovich Enman, was a military man. He was Estonian, born in October 1896 in Saint Petersburg. In 1913, he finished a secondary school in Kursk and enrolled in the Riga Polytechnic Institute. In August 1914, when World War I broke out, my father had just finished his first year of college. He decided to leave the Polytechnic Institute and volunteered to go to the front. He served first as a private and then as a noncommissioned officer on the Southwestern Front. He was wounded twice. In May 1915, his superiors sent him to a military college in Kazan for accelerated military training. When my father returned to the front in September 1915, he was first put in charge of a platoon, then served as head of a company. He was only nineteen years old. In August 1916, he was seriously injured and taken prisoner of war. While he was in captivity in Austro-Hungary, he was conferred the rank of lieutenant for not abandoning his positions. This is how my father became an officer of the Tsarist army. When he was finally released, he returned to Russia, and in December 1918 enrolled in the Red Army.

Why did he join the Red Army?

He didn't really have much choice. The entire nation was in ferment. Some people supported the February Revolution; others supported the October Revolution. When my father returned to Russia in 1918, the Bolsheviks were already in power. He served in different units. During the Civil War, he fought against Makhno in Ukraine. He then fought against General Miller near Arkhangel'sk, during the English invasion. When my father returned from that war, he was still young, only twenty-five years old, but he already had a lot of experience. Nowadays, young people only start working at nineteen. But back then people grew up much faster. My generation also grew up fast.

In 1926, my father graduated from the Higher Officer Training School "Vystrel." He served as head of a battalion and taught musketry first at the School for Reserve Officers in Poltava, Ukraine, then at the Miasnikov First Military Pilot School at the Black Sea. In May 1932, my father was appointed head of the courses for command officers of the mechanized and motorized troops of the Red Army, and my family moved to Moscow. After these courses merged with the Red Army Academy of Mechanization and Motorization, my father became head of the Department of Firing Instruction at the Academy. In 1934, he also graduated, by correspondence, from the Frunze Military Academy. In 1935, he was transferred to the Automotive and Armor Directorate of the Red Army, where he became head of the Department of Firing Instruction.

After my father was appointed head of the Department of Firing Instruction, my family moved into one of the buildings occupied by the command staff of the People's Commissariat of Defense in Lefortovo. Several years later, my father was arrested in that same apartment. It was a relatively large apartment, which had two rooms and a storage room; the latter also served as a bedroom for my grandmother and me. My mom and dad slept in a small room. The bigger room was taken up by my three sisters. The apartment also had a kitchen, combined with a bathroom. We lived in one of the three buildings constructed before the Revolution. At the time, these were considered to be luxury apartments. We even had a telephone.

Can you tell me a bit more about your father's side of the family?

My grandfather, Karl Avgust Iakob Enman, came from Tartu. I know this because I have copies of my grandparents' passports. He was a poor landless peasant, turned into a skilled worker. He traveled all over the country, boring artesian wells. My father was born in Saint Petersburg, where my grandfather was working at the time. I made a special trip to Tallinn to visit an archive there. In this archive, I found a vestry book from Tartu with information about my grandparents. The vestry book didn't mention my grandparents' nationality, but it talked about their Lutheran creed.

And what about your mother's side of the family?

My mom also came from an ordinary family. She was born in Pavlograd in Ukraine. Her father was a churchwarden, which means he was literate. My mom graduated from a gymnasium. This was uncommon, especially in provincial towns. My parents met in Ukraine and got married in 1920. They had four children, born in 1922, 1923, 1925, and 1928. I was the youngest.

But going back to my father's arrest. . . . As I said, my father started working at the Automotive and Armor Directorate of the Red Army. In the fall of 1936, Marshal Tukhachevsky instructed them to develop new methods of tank shooting. Previously, our tanks would try to hit a target without stopping, which was inefficient. According to the new method, a tank would need to stop briefly, then aim and fire. This method proved superior during the war. Yet, in May 1937, Mikhail Tukhachevsky was arrested. The new method was called "sabotage." This was the beginning of the purges in the Red Army. In 1937–1938, most middle- and high-ranking commanding officers of the Automotive and Armor Directorate were arrested. My father was arrested on June 12, 1938, as part of the Tukhachevsky Affair.[3] The charges against him were quite standard: he was accused of being part of the "anti-Soviet military fascist conspiracy, plotting to overthrow Soviet power by means of subversive activities, sabotage, and planning terrorist acts against the leadership of the VKP (b) and the Soviet government." Here I am quoting from the verdict of the Military Board of the Supreme Court of the USSR from September 27, 1938. Dozens of thousands of our citizens received similar charges.

My father was found guilty and sentenced to the highest measure, which meant death. All his belongings were to be confiscated. He was executed the same day as his verdict was announced, on September 27, 1938. We now know that Stalin and the Politburo signed execution lists for hundreds of thousands of people. By the way, Sergei Korolev's name was on the same list as my father's.[4] Stalin signed it on September 25, authorizing the execution of seventy-four military specialists and engineers. Korolev was saved by a miracle. My father was executed by a firing squad. Yet for a long time, my family didn't know what had happened to him. After my father's arrest, we were told that he had been sentenced to "ten years without the right to correspond." Only in the early 1990s was I finally allowed to see his dossier. Even though my father was officially rehabilitated in 1956, I didn't have access to any of these materials until much, much later. When I was finally allowed to see my father's dossier in a reading room at Lubianka,[5] I decided to take notes. This is how I know about this.

I was only ten years old when my father war arrested. The arrest and the search took place at night, when I was asleep. I learned about it by accident: I was playing outside with local boys and one of them called me "a son of the enemy of the people." I cried when I heard it for the first time. After my father's arrest, my entire family was evicted from our two-room apartments in Lefortovo. We were squeezed into one room of a communal apartment located in the basement on a building in Presnya. They decided to put two families in one room. But my family alone consisted of six people: my mother, grandmother, three sisters and me. Fortunately, the other family found some other accommodations.

Do you remember if your parents talked about politics or Soviet power at home?

I can tell you this: we were afraid to talk about such things back then. My mom didn't discuss certain topics with us, including my father's arrest. She was afraid that we could let something slip by accident. My mom was a homemaker, and before my father's arrest, she stayed at home and took care of us children. My father's mom lived with us, too, for a while. She died during the winter of 1941–1942, when the Germans came close to Moscow. Elena, my eldest sister, born in 1922, had to drop out of college and start working, because we had nothing to eat.

There are several dozen people whose parents had been repressed currently living in Chernogolovka. Sometimes we get together, usually around New Year. I know Mezhov-Deglin,[6] for example. He told me that he also had seen his father's dossier. The dossier had his father's handwriting. He said that the handwriting was clear in the beginning, but it got barely legible toward the end of the dossier. They beat them pretty badly there. When I read my father's dossier, I found out that my father had been also accused of going to different military units and giving lectures to commanding officers about the new method of tank shooting. In other words, he was accused of doing his job. Ironically, the war proved that the new method was correct. In fact, if we had used it during the first several months of the war, our army would not have suffered such heavy losses in tank battles. Later, during the war, we started using indirect fire, and this is exactly what my father was working on. This was the main part of his accusation. There were no other charges against him. Perhaps this is why he was among the first people to be rehabilitated. In 1956, someone told my eldest sister that they [Soviet authorities] started rehabilitating people, and she sent an appeal to the military prosecutor's office. I could not have gotten a job in Chernogolovka without my father's rehabilitation. You see, I graduated from the institute in 1955, but I could not get the security clearance necessary for working at a research laboratory.

Was it because of your father's arrest?

Yes, of course, it was. In fact, I only managed to attend college, because I concealed the fact that my father had been repressed. In a questionnaire, I simply wrote that he had died in 1938. I also left Moscow, following a friend's suggestion. It so happened that I finished five grades before the war. Then in 1941–1942 schools closed in Moscow, because the Germans came very close to the capital. As a result, I lost the entire year. When I started sixth grade in 1942, it was so cold inside our school building that we had to study in our winter coats. We were always hungry. I remember all we could think about at school were little bread rolls that our teachers distributed for lunch every day. Finally, in the fall of 1943, a policeman came to our apartment, asking if I was interested in going to a vocational school. Since many workers left for the front, they didn't have enough people working at factories. I had just turned fifteen years old, so I decided to drop out of school and started working at an auto-repair factory. In January 1944, I transferred to a factory located closer to my home and began working as a locksmith apprentice. This was the first job in my employment record book (*trudovaia knizhka*).[7]

What else do I remember from my childhood? When I was a boy and my family still lived in the basement, we got a letter from prison. It was addressed to the other family which was supposed to share a room with us. A member of their family apparently also had been repressed. But he survived and sent this letter. Somehow it was delivered to our address and ended up on our desk, and so I read it. The letter was addressed to Voroshilov. It informed him about the unlawful methods of interrogation in prisons. The author wrote that interrogators didn't allow prisoners to drink or eat, that they fed them salted herring instead, made them kneel on a floor covered with peas, and that they only allowed people to drink if they signed incriminating papers. Later, I found out that Voroshilov himself signed execution orders, together with Stalin, Molotov, and Kaganovich. At the time, though, Voroshilov was head of the People's Commissariat of Defense.

In general, I don't remember my childhood so well. I only have a few memories about my school. Maybe this is because I barely went to school. After I lost two years during the war, in 1944–1945, I enrolled in the school for working youth (*shkola rabochei molodezhi*). I barely attended it, but it allowed me to receive a certificate that said I had completed seven grades of school. Yet I always had a passion for education, so I decided to apply to college. I quit my job and went to Baku. My plan was to enroll in the Preparatory Military and Naval School (*Podgotovitel'noe voenno-morskoe uchilishche*) there. Its diploma gave you the right to apply to naval academies, and I hoped to attend college that way. To get there, I spent eight or nine days on a freight train to Baku. It took me so long because our train had to let all other trains pass first. The year was 1945, and I remember that trip vividly. I saw many destroyed villages, towns, and railway stations on my way there. I saw a lot of poverty-stricken people, walking barefoot along the rail tracks, and begging for food. The entire country lay in ruins. When we finally arrived in Baku, I found out that there might be a credentials committee present at the entrance exams. I got scared and decided not to apply. I was afraid that if they learned about my father, they could put in me in jail, too. How could I know they wouldn't? So I turned around and went back to Moscow.

In 1946, I applied to Riga River College and was admitted. I can tell you how this happened. I was eighteen years old, but I had not even finished my secondary education. My sister's former classmate, Vladimir Gruzdup, recommended that I go to Riga. Vladimir also lost his father during the repressions. But his parents got divorced just before the arrest in order to spare his family from repressions. Vladimir was Latvian. During the war, he fought in the Latvian Division of the Red Army. After the war, they allowed him to stay in Riga. He was a secretary of the Komsomol organization at the regional committee of the Communist Party. It was Vladimir who suggested that I go to Riga. He said they had opened new technical secondary schools there, which didn't have any credentials committees. So I went to Riga. I passed my entrance exams without any problem and was admitted to college. Later Gruzdup found me there, and recommended that I apply for a party membership.

Did you apply?

No, I didn't. When I was already in college, my company's commander, Major Aliev, an Azerbaijani, also tried to persuade me to join the ranks of the party. I was a good student and an athlete. But I always found excuses to avoid submitting my application. I said I was not ready. In fact, I didn't want to join the ranks of the party partially because of my father. I only became a party member after I moved to Chernogolovka. I will tell you about it later.

Going back to 1946, I think I simply got lucky that they allowed me to enroll in Riga River College. I graduated with distinction in 1950. Since I studied at the Department of Marine Engineering, my qualification was "a marine mechanical engineer." But I always wanted to attend university. So in 1950, I passed entrance exams and enrolled in the Leningrad Institute of Marine Transport Engineers. I graduated from it in 1955. After my father was rehabilitated in 1956, I got a job at the Kaliningrad Branch of the Laboratory of Engines of the Soviet Academy of Sciences.[8] I worked there for three years.

I remember in 1959 I went to Moscow to visit my mom. She had just received a separate apartment in Cheremushki, a residential area on the outskirts of Moscow. By that time, she had lived in the same basement for twenty-one years. You can still find this apartment building in Moscow—Mantulinskaya Street, apartment building 24. Her old room had a tiny window; only now they laid brick over that window. My mom and my sisters had been on a waiting list for a separate apartment for four years. Eventually, they received a two-room apartment. But since I left in 1946, the authorities didn't allocate any space for me in their new apartment, which meant I could not live there. These were some of the first prefabricated apartment buildings, the so-called *khrushchevki*.[9] My mom's apartment was on the first floor. It had a small hallway, a kitchen, and a bathroom. It was not a good apartment. But since she had lived most of her life in a communal apartment in the basement, even that small apartment looked like a palace to her.

And what happened to me? I already told you I left Moscow for Riga in 1946. I lived in Riga for four years, then in Leningrad for five years. Afterward, I got a job at the Kaliningrad Department of the Laboratory of Engines of the Soviet Academy of Sciences. In 1959 Stechkin, who was the head of that Laboratory, a laureate of the Lenin Prize, and a member of the Soviet Academy of Sciences, decided to transfer me back to Moscow. However, I needed a registration (*propiska*) to live in Moscow.[10] I went to the Moscow city committee of the Communist Party, asking them to register me at my mom's apartment, but they declined. They said they would not register me because there was not enough space in my mom's apartment for both me and my wife. So I went to the Presidium of the Academy of Sciences and explained the situation to them. They directed me to Fedor Dubovitskii at the Institute of Chemical Physics, saying that he was looking to hire young professionals.

Vladimir Karlovich, before we move on to Chernogolovka I would like to ask you another question about your childhood. Who took care of your education, when you were a child?

No one really did. My mom was a saint: despite all she went through, she never said a bad word about anyone. Our communal apartment in the basement had four rooms. Our room was the first on the left. I cannot remember who lived in the fourth room, but the other two rooms were occupied by a janitor and his family and a porter and his wife, aunt Grusha. They were all good people, and never asked about our repressed father. I would even say that ordinary people tried to help us out. In 1941, when the Germans came close to Moscow, they allowed us to move into a vacant apartment upstairs for a while. Many people evacuated from Moscow, but we had nowhere to go. I remember October 16, 1941 well, when the Germans came really close to Moscow. I remember people were burning books by Marx and Lenin, as German airplanes were flying over the city. They were afraid that the Germans would capture Moscow, and so they burnt books that could get them in trouble.

Did you have any books at home?

My father had *the Soviet Military Encyclopedia*. When the arrests of the military personnel began, he would paint over the names of the repressed with ink. There was also a book by Voroshilov about Stalin's role during the Civil War, and another book about the October Revolution. It was a large book in red binding. Most of these books, however, got lost during one of our moves. Despite this, when I was a kid, I read a lot. I was lucky that way: there was a library on the upper floor of our apartment building. My sisters usually checked out books and brought them home, and I read whatever they read.

What exactly did you read?

I read Jack London, Mayne Reid, and of course, a lot of Russian literature. That library saved me. I frequently used a reading room there, too, so all librarians knew me well. I think they even knew about my family's background. Sometimes I did my homework there. I remember that place quite well. It was not far from the White House, near Krasnaya Presnya Park along the Moskva River. There was a sugar refinery across the street from our house, which employed many women from the countryside. During the war, the refinery closed temporarily, and these women were transferred to work at the factory that produced explosives. Their faces and their arms always looked yellowish—it was bad for their health to work there. At some point, there was an explosion at the factory. Since it happened between two shifts, many people died.

I meant to ask you, by the way: when your father was arrested did you believe that he was an "enemy of the people"?

I was too little. I only started thinking about it when I turned fifteen or sixteen years old. I began to reflect on it in a more serious way when I was already in college, although I understood some things even before that. I will tell you this: I told no one about my father. No one, not even my wife. I was afraid. It was my only chance to get an education.

Vladimir Karlovich, what did you want to be when you were little?

What did I want to be? I simply wanted to attend college. I was attracted to the navy, the sea. This is why I went to Baku to apply to the Preparatory Military and Naval School. This is why I enrolled in Riga River College.

Did you family support your decision?

I was very independent. I also forgot to tell you that during the war, food was really scarce: 300 g of bread, 200 g of vegetable oil or something like this. But a worker's ration card was better: 600 g of bread. We also got half a liter of vodka. In the summer of 1943, I went to Tumanovo, not far from Vyazma—it is about 300 km away from Moscow[11]—to exchange vodka for flour. Near Gzhatsk[12] our train was bombed from air by two German airplanes. They wounded our engine-driver, so we had to walk the rest of the way to Gzhatsk. We had to walk through mine fields. Mozhaysk and Gzhatsk lay in ruins, but Tumanovo stayed intact, since the Germans had to leave it quickly. They left storehouses with provisions behind, and the local population got access to some of these provisions. The rumor about this reached

Moscow, which is why I decided to go to Tumanovo to barter. However, the woman I bartered with deceived me: instead of giving me 8 kg of flour she gave me 3.5 kg of bran. I only found out about this when I returned home. My mom was not at home, but my neighbor saw what I brought and explained that it was bran, not flour. I was so angry.

Vladimir Karlovich, what do you remember about the atmosphere in Russia when you were a child?

My childhood was overshadowed by the war. Every day we listened to reports coming from the front. The Germans started bombing Moscow in late July 1941. Initially their airplanes had little lights on their wings. But after our fighter planes shot down several of their airplanes, the Germans started flying without any lights. As I said, my childhood was overshadowed by the war. How to survive, how to find something to eat—these were the main questions for us. What else do I remember from my childhood? It seemed to me that everyone thought the same way. They probably didn't. It always ends badly when people start thinking the same way.

But I only started thinking about these things when I was in college. By the way, we had excellent professors there. I owe a lot to them. We had good science, history, and math teachers. We had a good library, too. When Stalin turned seventy years old in 1949, all Soviet newspapers were full of congratulatory letters and addresses to Stalin from different collectives. This made me wonder, for the first time, about the purpose of all this. Even before that, I remember the Communist Party issued a decree, which denounced Zoshchenko and Akhmatova.[13] But I liked Zoshchenko's stories and Akhmatova's poetry. There was also an assault on the composer Shostakovich. I started questioning whether these party decrees were indeed correct. They went to great lengths to clutter our brains, of course. See for yourself: they made us read *The Short Course of the Communist Party* first at school, then in college, then at the institute.[14] On the other hand, when Stalin died we went to Moscow by train to attend his funeral. We had to stand the entire trip from Leningrad to Moscow, since all seats had been sold out.

I am surprised to hear that you went to Stalin's funeral?

Yes, I did. And so did many of my classmates. Well, how can I explain this to you? Our brains were cluttered up: every day we heard how great and irreplaceable Stalin was. We didn't know about Stalin's repressions yet. Although subconsciously I always wondered: How could my father be "an enemy of the people," if he had four children? It just could not be! Fortunately, we came to the funeral too late. A lot of people died in the stampede in Moscow.

But why did you decide to go to his funeral in the first place?

My entire group went, and I went with them.

Does it mean that some people loved Stalin?

I don't know if they loved him or not. But when he died, we all thought that everything would fall apart without him. This is just the way we were raised.

Vladimir Karlovich, what did you think about the official ideology of the Communist Party, which promised to build Communism in the Soviet Union?

I can tell you this: I was shaken in 1956, when my father was finally rehabilitated. I even cried. I felt so bad for my mom, who had raised four children in the basement and not even once said a bad word about my father. At the time, I lived in Kaliningrad. I got married in 1955. Yet I concealed the fact that my father had been repressed even from my wife. Only in 1956, when he was officially rehabilitated, did I tell her that.

After we moved to Chernogolovka, I started making decent money, especially in comparison to other people. My salary was 300 rubles a month, which was very good money at the time. Besides, I had different inventions, and we received bonuses for their practical applications. However, we could not buy anything at local stores. Even in the 1960s, when Chernogolovka was still small, there were not enough supplies here, and we had to travel to Moscow to buy basic goods. In 1960, my older daughter Alla was born. In 1963, we had Natasha, my younger daughter. But we could not find clothes for our children in stores: 70 percent of the country's economy was working toward military production.

This is probably why I never really believed in building Communism. I could not understand what exactly this slogan meant: "From each according to his ability, to each according to his needs." We had nothing to eat, but Khrushchev promised that we would catch up and surpass America. I could not understand how this could be. Despite this, I passed all my philosophy and political economy exams. During exams, I told them what they wanted to hear. I talked about "the bright future." I didn't have much choice really. I can also tell you this: several students at the Leningrad Institute of Marine Transport Engineers were repressed. I personally know about two such cases. One student decided to play a joke: he somehow got on the institute's radio and announced "Attention! The Voice of America is speaking." It wasn't even political, what he said, but he was immediately expelled from the institute. Another student decided to argue with a political economy professor, disputing some of the numbers she was giving us. This student was also expelled.

Vladimir Karlovich, you mentioned attending the Leningrad Institute of Marine Transport Engineers.

Yes, I graduated from Riga River College with distinction in 1950 and was admitted to the Leningrad Institute of Marine Transport Engineers. I don't remember if I had to take any exams. I don't think I did. At the Leningrad Institute, I studied at the Marine Mechanical Engineer Department. I was a student in good standing and graduated in 1955. I got married around the same time. My wife and I went to Kaliningrad, where I was offered a job at the Marine Engineering Plant. I worked there until 1956. Then I transferred to work at the Kaliningrad Branch of the Laboratory of Engines of the Soviet Academy of Sciences, directed by Boris Sergeevich Stechkin. He was also head of the Moscow Laboratory of Engines. Stechkin had been repressed in the 1930s. His first arrest was part of the Industrial Party Affair. He was later awarded the Lenin Prize for his contribution to designing Tu-104 aircraft. I worked at the Kaliningrad Branch from 1956 to 1959. First, I was a senior engineer. Then the director asked me to organize a model shop. Our laboratory was located inside a shipbuilding plant which had been previously owned by the Germans. The plant had a special building for testing engines, and this is exactly where our laboratory was situated. I was promoted to main engineer, although later I asked to transfer me back to the scientific department, where I could conduct research. We tested various engines, mostly for submarines.

I already told you how I got a job in Chernogolovka. I first arrived in Chernogolovka on March 17, 1959, and officially became an employee on March 25. After I started working at the Branch of the Institute of Chemical Physics in Chernogolovka, Fedor Ivanovich [Dubovitskii] put me in charge of the institute's workshop, and then the Construction Bureau. At the time, only a few people worked here: three engineers and a couple of workers. Most people were still employed in Moscow and traveled there on a regular basis. In May 1959, the first twelve-apartment building was finished. My wife and I received two rooms in a communal apartment in that building. After we had our first child, we got a separate two-room apartment. When our second child was born, we moved into a three-room apartment. I currently live in a five-room apartment.

It took us two years to organize a machine workshop, which employed sixty people. We then opened the Construction Bureau, which at its peak employed between 200 and 250 people. In 1961, I was appointed the main engineer of the institute, which meant I was put in charge of both the machine workshop and the Construction Bureau.

Do you remember what Chernogolovka looked like when you first moved here?

The entire settlement consisted of a twelve-apartment residential building. There was also a two-storied hotel, where my wife and I lived for the first two months. Shortly after we moved here, they finished a second residential building, and my family move into it.

At first, there were no grocery stores in Chernogolovka. We had to bring everything from Moscow. I remember there was a canteen for construction workers, located in one of the barracks. We could buy rye bread there. White bread was rare, though. There was also a tiny grocery store in the village nearby, where one could buy things like salt. Yet to buy milk we had to go to the Chapaev sovkhoz.[15] But even that was complicated: to buy milk we had to contact employees at the sovkhoz, pay for milk ahead of time, and then pick it up in the evening. It was a big hassle.

What was the Chapaev sovkhoz and where was it located?

It was a huge state farm, which included different villages such as Chernogolovka, Iamkino, Botovo, Voznessenskoe, and others. As I already mentioned, there were no grocery stores here at first—only a canteen for construction workers. Another complication was that buses didn't really run to Chernogolovka at the time. They usually passed the town on their way from Noginsk, but there was no direct road connecting Chernogolovka to Moscow. The entire settlement was surrounded by forests and had minimal infrastructure. So whenever I went to Moscow for work, I tried to buy basic food supplies and bring them home. It was inconvenient, of course.

I am a bit surprised to hear this. I was under the impression that Chernogolovka was well supplied, at least in the beginning.

Definitely not in the beginning. I know this for sure because from 1961 onward I served as main engineer at the ICP Branch. In 1964–1965, I also was one of Dubovitskii's deputy directors. In 1962–1963, Fedor Ivanovich organized a grocery store in the Polymer building of the institute. Sometimes, he would go himself to different distribution centers in and around Moscow to arrange for provisions of food supplies. At one point, I remember, we could even buy caviar at our institute's store, right from the barrels. Red caviar cost 5 rubles per kilogram, black caviar was a bit more expensive. Even though we didn't have much money—an average salary of a junior researcher was 105 rubles—we could still afford it.

At the very beginning, in the early 1960s, we also had garden plots. They were located across the street from the two-storied hotel. We grew potatoes, carrots, tomatoes, and cucumbers there. We even planted an apple tree. I was shocked to find out, though, that employees of the local state farm didn't grow either cucumbers or tomatoes. Sometimes, they would come to us to buy fresh vegetables. It didn't last for a long time, however. In 1962, when they decided to transform Chernogolovka into a scientific town, that land was allocated for housing construction. Even though we had planted trees there, we had to give up that land. In 1963, they started building four-storied houses there—the so-called *khrushchevki*.

Later Dubovtskii got permission to build a large grocery store, "Gastronom," in Chernogolovka. When it finally opened, he invited the head of Mosobltorg, a distribution center in the Moscow region, and organized a reception in his honor. I didn't want to attend, but Dubovitskii told me I had to, since I was going to help out with "Gastronom" anyways. At the time, there were no other research institutes in Chernogolovka, and the ICP

Branch oversaw and maintained all the facilities in town. My department, for example, worked on various installations, including refrigerators, meat storage chambers, and other things. Dubovitskii managed to get funding for all of this.

How did he get funding? Did the Academy of Sciences help?

No, he traveled to Mosobltorg directly. There was also Noginsk Torg, and Dubovitskii went there as well, although they had stricter rules. Since there were constant food shortages in the country, Dubovitskii had to negotiate with various trade authorities. I was present at some of these conversations. For example, Dubovitskii would call up one of the centers for food distribution and say: "I am a deputy director to Nikolai Semenov." And Semenov was not only a member of the Soviet Academy of Sciences. He was also a vice president of the Academy and a candidate member of the Central Committee of the Communist Party. Besides, Semenov was a Nobel laureate, so everyone knew him. Since the party made all decisions in the country, Dubovitskii would start by calling a regional committee of the party. They would refer him to Obltorg, where he negotiated supplying Chernogolovka. All this was completely his initiative. He did a great job in this respect. This became especially important after they started building new research institutes in Chernogolovka in the mid-1960s.

And what about the grocery store that opened in 1962–1963? Was it supplied well?

Well, you could buy meat and butter there, which were both hard to get in the Soviet Union. As I already mentioned, they treated us to red caviar once in a while. But overall, there was a shortage there as well.

Could you buy vegetables in that store, for example?

No, I don't think so. What I remember, though, is that every year scientists and engineers had to volunteer at the Chapaev sovkhoz. We would help them harvest potatoes and, in return, they gave us a share of their crop. We even built a special facility to store vegetables, located on the territory of our institute. We helped them harvest potatoes for many years.

How exactly did this arrangement work? Did your institute negotiate a contract with sovkhoz?

Our institute's party organization was subordinate to the October regional party committee in Moscow.[16] So they would call up Dubovitskii and tell him to allocate several hundred people to work at the state farm. Usually, we had to stop working, right in the middle of our workday, and go help out at sovkhoz.

What did you think of it?

We thought it was necessary. Otherwise, we would not have had any potatoes at all.

You mentioned they gave you a share of potatoes you harvested, is that so?

Well, not exactly. At first, everything had to go to the state. Then the state would decide what share of the harvest we would get. Another problem was that the state farm often didn't have enough storage facilities. Sometimes, they stored potatoes right in the middle of a field: they put them in big piles and covered them with hay. Since storage was not properly organized, it was common for potatoes to rot.

Did heads of laboratories also work in the field, harvesting potatoes?

Yes, usually the entire laboratory had to take a day or several days off work, including a lab director, to "work on potatoes" (*rabotat' na kartoshke*).

Did the situation improve after they opened a large grocery store in town?

I would say so. One could buy milk, meat, and vegetables there. It was a big store, and Dubovitskii worked hard to organize food supplies.

Did this last long?

I would say it lasted until the 1980s, but the supplies came in waves: sometimes there was food in the store, other times there wasn't. Of course, Moscow was supplied much better. Yet Chernogolovka had better food supplies than provincial Soviet towns. Later, we also organized a rationed food parcel system,[17] so we no longer had to bring food from Moscow.

Was there a farmer's market in Chernogolovka at some point?

Yes, but it was built much later. In 1959, when I first moved to Chernogolovka, there was nothing here. The market was built in the mid-1960s, and our institute helped with its construction.

You mentioned earlier that after you moved here, you received an apartment almost right away.

Yes, for a while it was easy to get an apartment in Chernogolovka. In May 1959, shortly after we moved here, my wife and I got two rooms in a communal apartment. This was rare: since we didn't have children yet, they could have given us a room in a dormitory instead. In March 1960, when our first daughter, Alla, was born, we received a separate two-room apartment. It was a decent apartment: it had one large room, a smaller room, and a good kitchen. Residents with children got priority when apartments were distributed.

How typical was it that a young person like you received a separate apartment within a year after your started your new job?

I personally don't know of any other place in the USSR that offered such luxury. You see, previously I lived only in communal apartments. In 1938, when my father was arrested, my family got evicted. We had to move from our two-room apartment into one room of a communal apartment in the basement. I remember this room well: it had a tiny window and a 1.65-meter-high ceiling. When I grew up I always hit my head against that ceiling. I lived there from 1938 to 1946, when I left for Riga. But my mom and my sisters stayed there until 1959. The entire communal apartment had four rooms, and we had to share a kitchen and a bathroom with other residents. The rooms were dark, cold, and had tiny windows. On top of that, the window in our room faced a large stairwell leading up to the library upstairs. Because of that we had even less light inside our room.

Did you have electricity at least?

Yes, we had electricity. We also had oil-stoves (*kerosinki*), located in the narrow hallway of the communal apartment. As I already mentioned, originally my family was supposed to share a room with another family, which consisted of a mother and a daughter. But they went to live with their relatives instead. So it was just the six of us in one room. I don't even remember if I had a place to sleep in that room. There was certainly not enough space there for six people. This is why when I received my apartment in Chernogolovka, or my apartment in Kaliningrad prior to that, they both looked like a palace to me.

Did your first two rooms in a communal apartment in Chernogolovka have all modern conveniences?

Well, of course. These were very good apartments, which had gas, electricity, and running water. I think that some apartment buildings in Chernogolovka were even better than many apartment buildings in Moscow. They had high ceilings with ornaments, about 3 meters high, and wooden floors. They also had decent-size kitchens. Besides, all apartments were move-in ready. We stayed in that apartment until 1960, when my first daughter was born. Then in 1963–1964, when our second daughter was born we moved into a three-room apartment on First Street. It was a wonderful apartment: it was really spacious, with three rooms and its own kitchen and bathroom.

Did the fact that you were the main engineer and later a deputy director of the institute help you get a better apartment?

I don't think so. Apartments were usually distributed based on how many children a family had.

Do you remember what the cultural life in Chernogolovka looked like in the 1960s?

Chernogolovka was built by soldiers, military men, who lived in wooden barracks. In 1960–1961 they started screening films in one of these barracks. Tickets cost only 20 kopecks, and we would go there occasionally. Fedor Ivanovich also organized trips to various Moscow theaters. He used the institute's bus for these purposes. He always tried to make sure his employees had opportunities to relax. Later, they also built the House of Scientists in Chernogolovka.

Who built the House of Scientists?

The Academy of Sciences allocated funding for that. Still Dubovitskii had to make all the necessary arrangements. In 1962, he was appointed the Authorized Representative of the Academy of Sciences in Chernogolovka, which meant he was put in charge of building the town. I remember Dubovitskii bringing top officials from the Academy here, including its President, Mstislav Keldysh.

Vladimir Karlovich, some of your colleagues called Chernogolovka "an oasis of intellectual freedom." Would you agree with them? Was there more intellectual freedom here than in other parts of the Soviet Union?

Yes, I would say so. But it is mostly true for the late 1960s and early 1970s, when we invited Vysotsky and Okudzhava to give performances here.

Did you participate in organizing these concerts?

No, I didn't participate in organizing them, but I went to all the concerts. If you want to know what I contributed to building the town, I can tell you about the volleyball courts. When I was still a student at the Leningrad Institute, I was on the institute's volleyball team. When I moved to Chernogolovka, however, I found out that there were no volleyball courts here. I asked a local forester to help me find a suitable location near Chernogolovka, and our workers helped us turn this ground into a volleyball court. We formed a volleyball team, and even participated in various regional competitions. Later, they built sports facilities near a local lake, which included two basketball courts and a volleyball court. We helped them with construction.

Who built those sports facilities then?

Well, the Academy of Sciences allocated money for their construction, and soldiers who built Chernogolovka prepared the grounds for basketball courts, a volleyball court, and gymnastics apparatus. Then employees of the local institutes volunteered their time to finish their construction.

Do you remember when the first school was built in Chernogolovka?

They started building it in 1959, and finished around 1961–1962. By the time my older daughter turned seven, the school had been functioning for a while.

What did you think about the quality of education?

I thought it was not bad at all. Fedor Ivanovich invested a lot of time into finding good teachers. He selected them one by one. They all had to interview at our institute. I remember that when the school opened, my department organized student workshops there. We helped teach disciplines like drawing. One of our institute's employee taught chemistry there.

Is it fair to say that overall you were content with the quality of education that your children were getting?

Yes, a student who wanted to learn could have received a decent education.

Vladimir Karlovich, you mentioned earlier that there was more intellectual freedom in Chernogolovka. Can you tell me a little bit more about this?

Intellectual freedom, you say . . . I will tell you this: everyone was afraid. After all, our institute was a restricted access enterprise, and Dubotvitskii had to make sure that we didn't say or do anything that could get him or the institute in trouble. In what respect did Chernogolovka have more freedom? We invited artists here who were not allowed to perform on the big stage. This was the extent of our freedom.

Moreover, when I moved here, I was thirty-one years old, but I still was not a member of the Communist Party. Some inner voice always stopped me. I even tried to avoid participating in public work—I was so afraid I could say something that might have gotten me in trouble. You should understand that this is just how we lived back then. Well, in 1967 Fedor Ivanovich invited me to his office, where Zolotoi, who was a deputy director in charge of construction, and Eremenko, the head of the institute's party organization, were already waiting. They told me that I had to join the ranks of the party, because I had eighteen subordinates who were party members. How could I supervise them, if I was not a member of the Party myself? So I didn't really have much choice. I joined the party several months later, although I didn't want to. As I told you already, some inner voice always stopped me. Before 1956, I concealed that my father had been repressed from everyone, including my wife. We had to fill out questionnaires, and one of the questions was about my father. Even after the rehabilitation I still kept writing in all official documents that my father had died a natural death. He died instead when they shot him dead on the day of his trial, September 27, 1938. The trial itself lasted for five minutes. Nowadays, many people fail to comprehend what kind of country we live in and what we did to our country during the last century. You can hear arguments that attempt to justify our history. We industrialized, they say. Yes, but how many people perished as a result? Millions. Many people claim that we had built a great country. To this I say: "If that country was so great, why did it fall apart?"

And why do you think it fell apart?

Why? Because the economic system didn't work. Additionally, we had no freedoms.

Vladimir Karlovich, before you came to Chernogolovka, you lived in Moscow, Riga, Leningrad, and Kalinigrad. How did Chernogolovka compare to all these other places?

Chernogolovka was different because of the kind of people who lived here, especially in the beginning. Even after I moved here, I continued to travel a lot for work, so I could compare. People who lived here were more educated. During the first decade, there was a special atmosphere in the town. For example, we didn't lock our apartment doors. Actually, when I came to study in Riga in 1946, I found out that people there didn't lock their apartments either. One could walk down the street and see containers full of butter and sour cream. Latvian peasants brought these containers to local stores for sale and just left them there, knowing that no one would steal them. This would change later, after the Soviets launched the collectivization campaign there in the late 1940s. I was there at the time, and I saw what happened to those people who opposed collectivization. They were exiled. We didn't really understand what was going on. We only found out about what happened after it was all done. This topic was taboo then.

Did you go abroad during the Soviet period?

I did.

When did you travel abroad for the first time?

I went abroad even before the Berlin Wall came down. I went to the German Democratic Republic in the late 1980s.

Did you travel earlier, in the 1960s and 1970s?

No, I didn't travel anywhere then. No one invited us anyways. In 1988, though, the Academy of Sciences of the GDR got in touch with the administration of our institute. At the time, we were developing disintegrators for microorganisms—they could be used to get artificial blood—and the East German Academy of Sciences wanted us to develop such a disintegrator for them. They contacted the Soviet Academy of Sciences in Moscow, which in turn contacted Dubovitskii. At the end of the day, we decided we needed to travel to the GDR to make sure they had the technology for our disintegrator to work properly. That was my first trip abroad. In 1989, I went there again. We got there one week before the fall of the Berlin War. Yet I hadn't noticed any signs of it coming.

What did I think about those trips? I remember my colleague and I went to a local grocery store. And we saw twenty different kinds of sausages there, twenty different kinds of cheese, and other produce. I turned to my colleague—he was standing right next to me—and I said that I was not going to any other grocery stores. There were constant food shortages in the Soviet Union. We had a system of rationed food parcels, and even then one could only get limited food. I know some of our institute's employees went abroad even before the 1980s, but only few did. They usually were allowed to travel only to the so-called friendly democratic countries.

And what did you and your colleagues think about the West in general?

When the events in Hungary, Czechoslovakia, and Poland took place, they [the authorities] concealed the truth from us.[18] We only knew what the Soviet newspapers told us about these events. Of course, we talked about them outside of work, in private conversations. But we tried to be very careful, because we were afraid. At the institute, we had several party meetings in support of the Soviet response to these events. Some people spoke up to approve them, but most of us stayed silent. Our access to information was limited: we only knew what we the Soviet media told us. Later, of course, people would say courageous things criticizing Soviet actions, but it happened much, much later. It is easy to be brave when nothing is at stake. I must say that at the time a lot of people spoke in support of the Soviet government.

I barely remember anyone protesting openly. Maybe Igor Shchegolev said something critical, I don't remember now.

Vladimir Karlovich, what do you know about the Liubarskii Affair?

I heard about it. But you see, I didn't know Liubarskii personally and didn't really overlap with him at work. He taught astronomy at a local school, but as I said I didn't know him personally. I know that Andrei Sakharov came to Liubarskii's trial to support him.

I have heard that some of your colleagues attended the trial. Did you go as well?

No. Honestly, I didn't even know about the trial at the time.

Did they keep this information secret?

I simply didn't know anyone who was involved in the trial.

Did you identify with the official ideology of the Communist Party at any point?

The official ideology was good: everyone should be equal and free. The problem was that it had nothing to do with reality. Yet, as I already mentioned, I went to Stalin's funeral when he died. Thank goodness, we were not allowed behind the fence. Otherwise, we would have been part of that meat grinder.

And how did you react to Khrushchev's report at the Twentieth Party Congress? Some documents I read in the archives suggest that people didn't fully believe what he said about Stalin.

That's because they knew nothing about what had happened under Stalin.

And you?

I could already have guessed, partially because of my father. But even apart from that, I observed several other incidents. When I studied at the Leningrad Institute, one of my classmates was expelled for making a joke on the institute's radio. Then in 1950 1951, our math professor Belen'kii was arrested. Even before that, at Riga River College two students were expelled. They came from Smolensk and complained about the poor living conditions for people who worked at collective farms there. Even though no one spoke about this openly, one could not help but notice all these things. Eventually, they left their mark.

Later on, dissidents attempted to speak about this openly, Andrei Sakharov, for example.

Well, Andrei Sakharov . . . you see it was different. They didn't arrest him for what he said, because of his privileged status. But if it were an ordinary person voicing these concerns, he would have been arrested long before.

Do you think Sakharov's dissident activity was beneficial for the country?

Of course, it was beneficial. Nowadays, anyone can read Solzhenitsyn's books, for example. In one of his essays, he writes about freedom, especially freedom of speech. He argues that we need to stop pretending to be "great," stop organizing parades, and so forth. Even when they talk about the Great Patriotic War, they tend to glorify what happened. But I lived through that war, and I saw it with my own eyes. I saw soldiers leaving to go to the front, and I saw them again when they returned.

Vladimir Karlovich, do you consider yourself to be a part of the so-called "thaw generation," or shestidesiatniki generation?

I am not sure how to answer this question. *Shestidesiatniki* are typically those who openly voiced their dissent. I mostly agreed with them, but I never dissented openly. I agreed with them silently. At least, I never preached what I didn't believe. I stayed silent. For example, we had meetings of the institute's party organization, but I never spoke up at those meetings. Only once do I remember speaking up. It happened around 1980. I remember there was a meeting of the party organization at the Institute of Chemical Physics in Moscow, which was attended by the representatives of the October regional party committee. Dubovitskii asked me to prepare a talk about the work of my department. And I told them about the deteriorating conditions for workers at my department: their low salaries and heavy workload. As a result, I argued, few people were willing to stay in the profession and the average age of employees was around fifty years old. After the meeting, Fedor Ivanovich asked me to step outside with him and told me that I went too far in my criticism. He was sitting next to a representative of the October party committee, and I guess that person said something to Dubovitskii.

And what was your reaction to Gorbachev's coming to power in 1985 and the launch of perestroika?

I took it positively, although I thought that Gorbachev spoke too much, but took little action.

Then what did you like about him?

Gorbachev was the first Soviet leader who started talking about the need to change the international situation. For example, he said that we needed to get rid of the Berlin Wall.

Does this mean you supported Gorbachev's foreign policy?

Yes, I was initially attracted to his foreign policy. He also began to revise the thesis about the leading role of the Party, and I could support that, too. You see, if you were not a party member in the Soviet Union, you could not really advance professionally. So many higher-up people at our institute had to join the ranks of the party. At some point, Dubovitskii told me directly that I was letting him down because I was not a party member. He put me in a leading position, despite this, but I guess people above him didn't approve of that. So I joined the party, but I always thought that the party's promise to make everyone equal and to satisfy everyone's needs was utopian. People are not equal. They never were, and they never will be.

NOTES

1. On how the Stalinist Terror affected the private lives of Soviet families, see Orlando Figes, *The Whisperers: Private Life in Stalin's Russia* (New York: Metropolitan Books, 2007).

2. See "The Victims of Political Repressions in the USSR" database, Memorial Society Archive: http://base.memo.ru/person/show/2659808.

3. Mikhail Tukhachevskii was a Marshal of the Soviet Union and in the 1930s played a central role in modernizing the Red Army. He was arrested in May 1937, by Stalin's order, accused of espionage and high treason, and executed in June 1937.

4. Sergei Korolev is considered to be the founder of the Soviet space program, which made possible the launch of Sputnik in 1957 and Yuri Gagarin's first flight into space in 1961. During the Cold War, Korolev was the chief designer of rocket technology in the Soviet Union.

5. Here Enman refers to the headquarters of the KGB on Lubianka Square in Moscow.

6. Leonid Mezhov-Deglin is an experimental physicist employed at the Institute of Solid State Physics in Chernogolovka.

7. Employment record books were introduced in the Soviet Union shortly after the Russian Revolution. It was an official document, which recorded an employment history of its owner.

8. The Laboratory of Engines had two branches: one in Moscow and one in Kaliningrad. In 1961, the Laboratory was reorganized into the Institute of Engines of the Soviet Academy of Sciences. From 1954 to 1962, the Laboratory was directed by Boris Stechkin (1891–1964), a Soviet scientist and engineer, who worked on the theory of heat engines and was involved in construction of different aircraft engines.

9. *Khrushchevki* were low-cost prefabricated apartment buildings, constructed across the Soviet Union during the Khrushchev era (1953–1964). They were often of low-quality, built quickly from concrete, and preassembled parts. Yet their construction allowed millions of Soviet citizens to get their own apartments for the first time in their lives and move out of communal apartments, or *kommunalki*.

10. The "propiska" system was the enforced obligatory registration to a residence, prevalent in the Soviet Union.

11. Here the interviewee provides a rough approximation. The distance between Moscow and Vyazma, in Smolensk region, is 234 km; the distance between Moscow and Tumanovo is 208 km.

12. Present-day Gagarin.

13. The denunciation of the satirist Mikhail Zoshchenko and the poet Anna Akhmatova in 1946 was part of the late Stalin regime's assault on the Soviet cultural intelligentsia.

14. *The Short Course on the History of the CPSU (b)* was first published in 1938. It was commissioned by Joseph Stalin and contained Stalin's own exposition of the fundamentals of Marxism-Leninism. It became the canonical text of Stalinism and was a required reading for every schoolchild or student in the Soviet Union.

15. Sovkhoz was a state-owned and state-operated farm in the Soviet Union.

16. Regional party committees of the CPSU overlooked party organizations on the ground and were subordinate to the Central Committee of the CPSU.

17. The system of rationed food parcels was quite common in the Soviet Union. It usually required a volunteer to make a list of required foodstuffs and place an order at a large grocery store in Moscow. A food parcel contained such basic food items as butter, cheese, meat, sausage, and buckwheat.

18. Here Enman is referring to the Hungarian uprising of 1956, the Soviet invasion of Czechoslovakia in 1968, and the Solidarity movement in Poland in 1980–1981.

Chapter 3

"After Winning Such a Devastating War, We Believed That We Could Do Anything"

Georgii Borisovich Manelis

At the age of thirty, Georgii Borisovich Manelis (1930–2015) found himself in charge of one of the first four research laboratories at the ICP Branch in Chernogolovka. This stunning promotion, as well as Manelis's professional success over the following decades, significantly shaped his positive outlook on Soviet society. When I conducted a series of interviews with him from 2009 to 2011, Manelis, a doctor of chemical sciences, was still a prominent figure in the Chernogolovka scientific community: he was a deputy director at the Institute of Problems of Chemical Physics and a corresponding member of the Russian Academy of Sciences. Older residents of the town referred to him as one of the "founding fathers." Perhaps it was this special status in the local community that prompted Georgii Borisovich to reflect on his generation more broadly when we met for a series of interviews in December 2009 and February 2011. Unlike the first two interviews in this volume, which focus on individual scientists' overcoming the traumas of wartime and Stalin's terror, this oral history paints a different, somewhat heroic image of the Soviet Union. It conveys the idealism of many young people who came of age in the wake of World War II. At the same time, the reader will notice Manelis's reluctance to discuss the repressive side of the Soviet system. In fact, he continued to romanticize this system even after the Soviet collapse.

The origins of Manelis's positive outlook on Soviet society can be traced back to his early years. Both of his parents belonged to the new Soviet intelligentsia—a product of the October 1917 Revolution. Manelis's mother came from a family of Ural Cossacks. Her parents died during World War I, leaving her to take care of four younger siblings. Despite this, in the 1920s she managed to attend university, finish graduate school in Moscow, and become a professor of philosophy. Manelis's father was born into a large

Jewish family, which resided in the Vinnytsia region of Ukraine. He went on to study in Moscow as well, becoming a professor of law. It was largely thanks to the Revolution, Manelis was convinced, that his parents could get a higher education, start a family, and have successful academic careers. Even so, he acknowledged that his parents only managed to survive the Great Terror because they had left Moscow and moved to Tashkent in the Uzbek Soviet Socialist Republic around 1929, where they accepted positions at the Central Asian State University. Georgii Borisovich grew up in Tashkent in the 1930s and 1940s. During the war, his family entertained famous Soviet artists and writers who had evacuated to the city. Later Manelis's mother was elected to be a Deputy Chairwoman of the Supreme Soviet of the Uzbek Soviet Socialist Republic,[1] which advanced his family's status among the new Soviet elite.

While many of his peers talked about their family's suffering during the Great Patriotic War, Manelis focused on the rise of patriotism. The impact of the war on shaping his generation is one of the recurring themes in this chapter. The Soviet victory in the war, Manelis argued, strengthened his generation's commitment to the Soviet regime. It convinced many of his peers that they had endless opportunities available to them. While this idealism was indeed widespread among the postwar Soviet youth, Manelis's unwillingness to discuss the human cost of Soviet victory—even decades later—is unusual for a highly educated member of the intelligentsia. Another important theme in his account is the intimate relationship that existed between scientists and the late Soviet state. Like many of his colleagues who entered the profession during the first Cold War decade, Georgii Borisovich highly valued the extraordinary professional opportunities available to him. His own meteoric rise from a graduate student to head of a laboratory—all within less than a decade—convinced him that the Soviet Union created some of the best conditions for the development of natural sciences.

Georgii Borisovich was not such a fervent Communist believer as the next interviewee in this volume. Nonetheless, he strongly sympathized with socialist ideology. His professional success and his faith in the transformative power of modern science made him a staunch supporter of the Soviet state. In the interview, Manelis admits many shortcomings of the Soviet economy, such as its inability to adequately feed people or to ensure continued economic growth. Yet he argues that the highly centralized Soviet system was best suited to solve "the big problems of science," like building nuclear weapons. Georgii Borisovich overlooks the fact that there were serious problems in this sphere as well. His reference to Valery Legasov appears to be sadly ironic in this context.[2] Manelis's discussion of Karl Marx's philosophy later on in the chapter suggests that he bought into many premises of Marxism as well. According to Manelis, the class struggle, described by Marx, had evolved into economic exploitation on a global scale in the early twenty-first century, led by the United States. While recognizing the failure of Communism "to take into account the biological needs of human beings," Georgii Borisovich points to the Chinese variety of Communism as a model of success.

Neither Manelis's trips to the West nor his encounter with Soviet dissidents seemed to undermine his commitment to the Soviet state. In the 1960s and 1970s, Georgii Borisovich was one of the few scientists in Chernogolovka who was allowed to travel abroad. His first trip was to the University of Cambridge in England in 1964, where he attended an international symposium on combustion. The fact that he was allowed to travel to a Western, capitalist country, without any previous experience going to Eastern Europe, was unusual. Even more striking was the fact that he was not even a member of the Communist Party at the time. Georgii Borisovich went to Czechoslovakia, France, and even the United States afterward. He never had a problem getting an "exit" visa, which was necessary to leave the USSR. He valued this privilege highly, which he saw as yet another sign of the Soviet state's recognition of the elite status of Soviet scientists. Manelis's encounters with Soviet dissidents had no significant impact on his worldview either. Like many members of the late Soviet intelligentsia, he read Aleksandr Solzhenitsyn's One Day in the Life of Ivan Denisovich, *when it was first published in* Novyi Mir *in 1962. But he didn't actively seek to read samizdat. In the interview, Manelis argues that he and many of his colleagues could sympathize with the dissidents' criticism of the Soviet regime, yet they thought that the dissidents' activity was overall futile.*

The collapse of the Soviet Union in 1991, and the decline of the scientific profession that followed, proved to be extremely traumatic for Georgii Borisovich. He held Mikhail Gorbachev accountable for the decline and believed that perestroika was a big mistake. While recognizing the need for reforms, Manelis maintains that Gorbachev should not have started political liberalization and should have focused on economic reforms instead, gradually improving the standard of living in the USSR. Toward the end of the interview, he is equally critical of the drastic economic liberalization that took place under Boris Yeltsyn in the early 1990s.

Let's talk about our generation for a bit. First, I remember that even before the war broke out industrialization had started changing the outlook of our country and the standard of living. We didn't hear about the Gulag when we were children, but we did hear about the Papanin expedition to the North Pole and Valerii Chkalov's first transpolar flight. These people were our heroes. Of course, propaganda played its role, but there was more to it than blatant propaganda. Second, the Great Patriotic War—our generation suffered a lot from the war. We were children of that war, if you will. Almost all grown men whom I knew were drafted. All the men in my family were drafted into the army, every single one of them. My family lived in Tashkent, and during the war many of our relatives came to stay with us, bringing their children with them. The war had a huge impact on my generation, especially the fact that we had been attacked. I remember June 22, 1941, very well, even though I was only eleven years old. Our spirits were high at first. "We'll defeat those Germans in no time!" we thought. But this mood didn't last long. By the winter of 1942, the real war set in, when the entire country worked to support the war effort, worked toward our victory.

We were children, but we took everything that was happening close to heart. Since we lived in Tashkent and our army needed clothing, we began picking cotton in the fields when we were in fifth grade. Many of my peers quit school altogether and started working. At some point, our new four-storied school building was transformed into a hospital for soldiers. We had to move to smaller buildings nearby, but it was natural for us. The war was underway, and these were our soldiers. Besides, everyone lived like this. Then, in 1942, our army started winning: the defense of Stalingrad, the heroism of Leningrad residents during the siege of the city. These were challenging years, but to us children, there was never a question about who would win the war. The thought of losing never even crossed our minds.

And the year 1945—do you know what this year meant to us? After all these devastation and suffering—Victory! Our troops were in Berlin, Hitler was defeated, and we liberated half of Europe. This was extremely exciting. Almost all of Europe, except for Britain, was occupied by the Germans and was fighting against us. Of course, Britain fought hard to defend itself against German bombings. But none of these could compare to the Battle of Moscow or the German bombings of Stalingrad. And it was not just the victory for our troops: most factories in the Soviet Union worked to support the war effort. This is why when we won it made us feel enormously proud. This intense feeling of pride and the realization that we won and that "we could do this" had a tremendous impact on the formation of our generation.

Another important thing that I want to talk about is that right after the war there began a quick and harsh process of reconstruction. It didn't seem harsh at the time: after the war, everyone lived in horrible conditions and ate poorly. But overall the country was recovering. All over the place, they were building new factories and housing. Soon they abolished ration cards and made bread and sausages available in grocery stores. Moreover, the Soviet government reduced prices every year. So again, we felt like we had overcome these hardships as well. We—mostly our parents, but us too—we made it possible! This became the leitmotif of our generation. We believed that if we worked hard enough, things would work out. And things did work out! In 1945, the Americans exploded their first nuclear bomb, but by 1948, we had the bomb, too.[3] In 1953, we built the first nuclear electric station—the first one in the world. We had the first passenger jet and many other things that I won't mention now. After winning such a devastating war, we believed that we could do anything. Of course, the Doctors' Plot in 1953 made us feel uneasy, but most of my peers thought it was some sort of aberration, which had little to do with all our accomplishments.

And there is one other circumstance that I want to mention. I already told you that after winning the war we believed that we could do anything. But there was also a widespread belief at the time that "science could do anything." And it was not just true for our country, but for the rest of the world as well. Nuclear energy held a lot of promise around the globe. Then there were antibiotics and polymers that had been widely used during the war. So the postwar Soviet youth largely embraced these two notions: "we can do anything" and "science can do anything." To this day, I tell my graduate students that there is nothing that they can't do if only they work hard enough. Later in the 1970s and 1980s, there came a widespread disillusionment with the transformative promise of modern science. But back in the 1940s and 1950s, we believed that science could do anything, that it could help us establish paradise on Earth. Such views were not unique to the Soviet Union. They were also common in Europe and in the United States.

Georgii Borisovich, you are saying that the war, and especially the Soviet victory in that war, spurred a lot of enthusiasm among Soviet people, in a sense contributing to their loyalty to the Soviet regime. Yet, there are numerous memoirs in which participants recount how the war led to their disillusionment with Soviet power.

I guess this could have been true for the war veterans, who crossed the Soviet border and went all the way to Berlin. I cannot speak for them. I lived in Tashkent, and a lot of my friends and relatives lived behind the frontlines.

Did you know, for example, that some Soviet soldiers were sent to the Gulag immediately after the war?

We knew about the camps, but we didn't know that people were sent to the Gulag right from the frontlines. I think most of these people were criminals, not political prisoners. Some of them violated the rules or committed a crime. It's hard for me to tell. But I did know about the camps starting in 1937.

What exactly did you know?

I knew that they existed. I also knew that my uncle had been executed. Not even that, I just knew that in 1937 "he had been taken." We didn't really discuss it in my family.

And what did you think about it?

We were children, and my parents didn't talk about it with us. But you see, none of this had a major impact on our formation and our views. In 1944, when we turned fourteen years old, for example, we all joined the Communist Youth League. No one forced us, but the war was underway, and we could not have imagined acting otherwise. "Communists, forward! Komsomol members, forward!" We simply *had* to be part of the Komsomol. This was just the way we felt. I am talking about ordinary people now. Perhaps there were people who were more aware of the repressive side of the Soviet system. But the vast majority of the people focused on our victory in the war and the promises of modern science.

Why am I telling you about all this? It is against this background that Chernogolovka was created. It all started with the Institute of Chemical Physics in Moscow. Nikolai Semenov, the director of the Institute, decided to build a testing ground in Chernogolovka. The ICP had been involved in research on explosives for a while, and in the early 1950s, it started conducting research on new powerful explosives. It was challenging to run tests on explosives in Moscow, however, so they decided to establish a testing ground outside of the capital. From the very beginning, Semenov was planning to expand the research conducted in Chernogolovka beyond the testing ground, which is why they started building lab facilities here right away. The decision was made in 1956. The only problem was that at the time there was nothing in Chernogolovka except for a tiny village surrounded by marshes and forest. Since there was no infrastructure here at all, employees of the Moscow ICP didn't want to move to Chernogolovka. I remember one of my peers telling me that he could write and publish four articles, while I was draining marshes and uprooting trees. But we were young—we were in our mid-to-late twenties—and full of enthusiasm, so we went.

At first, we had to spend a lot of time taking care of construction and administrative issues. We helped design lab facilities, for example. But as I said we were young and full of energy and ideas, and Semenov decided to take a chance. When in 1960 he created the first four laboratories in Chernogolovka, he put us in charge. We were only thirty years old. This was extremely unusual, since the average age of a lab director was around forty-five years. Semenov had to overcome significant resistance from his colleagues at the Presidium of the Soviet Academy of Sciences, which had to approve of our appointments. I am telling you all this to demonstrate the general mood in the Soviet Union at the time. There's a marsh—so what? We will drain it and build a town in its place. This belief that we could do anything was the leitmotif of our generation.

Semenov didn't share his plans with us right away. But as we were building a testing ground, he was already planning to expand the branch of his institute in Chernogolovka. When our enthusiasm and euphoria started yielding results, he entrusted us with developing new directions of scientific research. We didn't just work on some narrow themes, but we developed entirely new areas of science. For example, some of us studied detonation, others researched combustion. I conducted research in the field of kinetics for a while. We were young and bold, and we didn't shy away from any challenges.

Our original task was to research new powerful explosives. Soon, however, our focus shifted to solid rocket propellant, since the state needed propellant for intercontinental ballistic missiles. Semenov's testing ground in Chernogolovka became one of the centers that worked on that. He told us to focus on the theory and to leave engineering tasks to others, so we worked on theoretical issues. Later, Semenov also invited biologists and biochemists to come to the ICP Branch in Chernogolovka. His ultimate goal was to build a scientific town, populate it with young people and give them the freedom to do science. Now I understand that he was taking a huge risk, asking the state to invest a lot of money into this project. Besides, as I already mentioned, he had to overcome the resistance of the Presidium of the Academy of Sciences. There were many people at the Academy who were opposed to giving so much freedom to a group of youngsters. They called us "Semenov's kindergarten." But it all worked out in the end.

Once Semenov got his project off the ground, he started expanding the testing ground by creating new laboratories and recruiting more scientists. In the early 1960s, he helped establish another institute in Chernogolovka: the Institute of Solid State Physics. They recruited many young scientists there as well. A lot of them came from the Institute of Physical Problems in Moscow, directed by Petr Kapitsa. You probably know that Kapitsa was Semenov's close friend. They were friends since they worked together at the Leningrad Physico-Technical Institute in the 1920s.[4] Only later did we find out that Semenov was planning to create a scientific center at the intersection of physics and chemistry. He called it "Russian Cambridge." Several years later, the entire school of theoretical physicists of Lev Landau moved to Chernogolovka. They established the Institute of Theoretical Physics here. So gradually Semenov's idea started to come to life. We all lived and worked together, and again we felt that we could do anything.

Georgii Borisovich, let's come back to this in a little bit. For now, I would like to talk more about your childhood. I know that you were born in Tashkent in 1930. How did your family end up in Tashkent?

My parents moved to Tashkent from Moscow in 1929–1930. They both finished graduate school in Moscow and were assigned jobs at Central Asian State University in Tashkent. At the time, Tashkent was the cultural hub of Central Asia. It attracted many members of the Soviet intelligentsia. During the war, a lot of artists and writers evacuated to Tashkent, including Kornei Chukovsky, Anna Akhmatova, and Alexei Tolstoy. My parents knew many of these people.

My mom was born in 1897, and came from a family of Ural Cossacks. Her father went missing during World War I. Her mother also died early—either during World War I or the Civil War. My mom was left alone with four younger siblings: three brothers and one sister. Even though she was a child herself, she had to raise them all by herself. Her uncle helped her some, although I don't know much about him. All I know is that he was a part of the Bolshevik underground before the Revolution and that he became a professor of economics afterwards. In 1937, my mom's uncle was arrested and executed, together with my mom's eldest brother. They were both accused of trying to establish an independent Ural Cossack state or something like that. I'm not sure: my parents didn't tell us much about it.

Since my mom always dreamed of attending university, she took her three younger brothers and went to Saratov, where she was admitted to Saratov University. While she was at school, she arranged for her brothers to stay at an orphanage, but she visited them all the time. When they grew up, they were able to get educated as well: one of her brothers became a pilot, another one was a railroad worker. In 1941, they were all drafted, but none of them was killed at war. I first met my mom's relatives when they came to live with us in Tashkent during the war. As for my dad, he was born in the Vinnytsia region of Ukraine. His father died in the 1930s, and my dad was raised by his grandmother. He came from a large family, too. My dad studied in Moscow. It was also in Moscow that he met my mom, although they didn't register their marriage for a while, since they were both progressive young people. Only after the war did they officially get married.

My mom was a professor of philosophy. She knew many physicists, chemists, biologists and was widely respected by her colleagues. We had a lot of books on natural science at home—Lamarck, Darwin, Einstein, and many others. My dad was a lawyer, specializing in the theory of the state. He introduced the concept of restricted sovereignty. Both of my parents only managed to avoid the repressions of the 1930s because they left Moscow and moved to Tashkent: out of sight, out of mind, as they say. There were purges in Tashkent, too, but they mostly targeted local nationalists. Later in life, my mom was elected to be a Deputy Chairwoman of the Supreme Soviet of the Uzbek Soviet Socialist Republic. She was elected because at the time they had strict guidelines about the ethnic and class composition of the Supreme Soviet: it was mandatory that a chairman of the Supreme Soviet was an Uzbek and a deputy chairman was a Russian. The Soviet also had to represent different layers of society: workers, peasants, and the intelligentsia. Besides, in Central Asia they had to make sure that they had a certain percentage of women in their representative body. My mom fulfilled many of these requirements.

As I already mentioned, during the war Akhmatova, Chukovsky, Tolstoy, and other prominent people evacuated to Tashkent. I knew them personally through my parents. Our apartment was extremely cramped: my father's sisters and my mother's sisters-in-law came to live with us. As a result, we didn't have much space to entertain guests. By the way, my father was drafted in 1941. By the end of the war, he was promoted to major. They wanted him to stay in the military, but he got demobilized. After the war, he became a department chair at Central Asian State University.

What role did the Russian Revolution play in your parents' lives?

Well, it allowed both of my parents to have successful academic careers. My mom came from a family of Ural Cossacks. My dad's family was Jewish; they lived behind the Pale of Settlement before 1917. The Revolution allowed both of my parents to go study in Moscow. It was also the Revolution that made possible a marriage between a Jew and a daughter of an Ural Cossack. In this respect, mine was a typical Soviet family.

I graduated from the university in Tashkent in 1953 and decided to apply to graduate school in Moscow. My wife, Nina Petrovna, graduated from the Medical Institute. We met in high school and decided to get married when I was a senior at the university. Nina had to study for one more year to get her degree. As soon as she graduated, she came to live with me. We rented apartments in Moscow and the Moscow region for about seven years, moving from one place to another. Then Emanuel invited her to work in his laboratory, and she decided to switch to an academic career. Nowadays, she is head of a laboratory at our institute. She is a doctor of biological sciences. Overall, a lot of scientists who moved to Chernogolovka came from remote parts of the Soviet Union: we came from the Uzbek SSR, some of my colleagues came from the Far East, Rostov-on-Don, Belarus, and many other places.

And why did you move to Moscow in the first place?

In 1953 I was admitted to graduate school at the Institute of Chemical Physics.

Who was your dissertation advisor there?

Nikolai Aleksandrovich Chirkov.

Wasn't he Dubovitskii's friend?

Exactly. They were both from the same village and went to school together. In fact, Dubovitskii came to the ICP shortly after me, around 1954. You see, I graduated from the university in 1953. When I began my studies in 1948, it was only three years after the war, and few people could predict what kind of specialists we might need in the early 1950s. By the time I got my diploma, there was a growing demand for scientists across the country. Science was on the rise, but there were not enough specialists. After we graduated, they sent us to continue our studies in graduate schools in Moscow. I remember they delegated twenty people from the Uzbek SSR. Half of them were Uzbeks. They were admitted to graduate school to fulfill quotas. They were admitted on the following condition: after graduation they had to return to the Uzbek SSR to develop science there. By contrast, I didn't fill any quotas to get into graduate school, so I had to take entrance exams with everyone else. I arrived in Moscow in September 1953 and started looking at different graduate programs. I applied to the Mendeleev Institute of Chemical Technology first. I remember I passed my chemistry and language exams, but failed my exam on Marxism-Leninism. I was naive and somewhat bold and I had quite liberal views so to speak.

How would you describe your "liberal views"?

Well, I liked Bukharin's program which called for the continuation of the New Economic Policy.[5] The professor who gave the exam, by contrast, was an old school Marxist-Leninist. He accused me of "Bukharinist leaning," and I called him a Trotskyist, since he supported the idea of a world Communist revolution. He threw me out of the room, of course. Since I came to Moscow through the Academy of Sciences, I went to the Presidium of the Academy and asked them for help. Some young girls in the graduate school office recommended that I apply to the graduate school at the Institute of Chemical Physics, since "they admitted everyone." I decided to take their advice and was able to transfer my grades from the Mendeleev Institute to the ICP. I passed my ideological exam without much trouble. This is how I became a graduate student at Semenov's institute.

Did you not know that you could only express a specific viewpoint during your exam?

I certainly knew that. We all had to read Stalin's *Short Course on the History of the Communist Party*. But keep in mind that ideological slogans were quite different from real life. You probably have heard, for example, that on paper Stalin's Constitution of 1936 was the most democratic constitution in the world, but it had nothing to do with reality. These exams were not that different.

So in 1953 you started graduate school at the ICP. Did you live and work in Moscow for the next three years?

Yes, and by the way, there was not much connection between the research I conducted in graduate school and my work in Chernogolovka.

Did you defend your dissertation before you moved to Chernogolovka?

Yes. In 1956, when they decided to expand the ICP, I was finishing up graduate school. I could not stay in Moscow. For one thing, I didn't have a Moscow registration (*propiska*). For another, I came from a Jewish background. And even though I wrote "Russian," in response to a question about my nationality in various forms, my last name still gave me away. Moreover, I came from Uzbeskistan. Different top scientists at the ICP, including Emanuel, Chirkov, and Voevodskii, requested permission to keep me employed at the ICP. But the political department of the institute said "no." Now, at the time, there was a political department instead of a party organization at the institute, since the ICP had joint projects with the Ministry of Medium Machine-Building.[6] Later, we would have a decent party organization, but back in 1956 I still had to deal with the political department. It was an extremely reactionary organization, which had broad authority. It would be abolished in 1957.

Anyways, as I already mentioned, Chirkov wanted me to keep working in his laboratory in Moscow, but it was impossible. So Dubovitskii, who also worked in Chirkov's laboratory researching explosives, invited me to come work with him at the ICP testing ground in Chernogolovka. The political department could not care less about Chernogolovka, since there was nothing there at the time except for forests and marshes. They simply didn't want me to stay in Moscow. They kept asking me why I didn't want to go back to the Uzbek SSR and develop chemical physics there.

And why did you not want to go back to the Uzbek SSR?

When I started graduate school at the ICP, I found myself in the very thick of events. I was surrounded by great men, like Semenov, Zel'dovich, and Khariton. Science was on the rise in the Soviet Union, and the atmosphere at the institute was especially creative. Besides, my housing situation was complicated. We already had a daughter but had nowhere to live. No one promised to give us housing, but we believed that everything would come together eventually. Young people were more modest back then than they are today. So when Dubovitskii offered me to come and work in Chernogolovka, I jumped at that opportunity.

What exactly did Dubovitskii offer you?

He said that if I came with him, I would study combustion and explosives. In graduate school, I conducted research on polymers, but this new area of research looked equally exciting.

Did you move to Chernogolovka right away?

No. At first, we worked at the ICP in Moscow. The resolution to establish a testing ground had just come out, but the construction of Chernogolovka was still in the initial stages. We worked at the Moscow ICP until 1960. It was a busy time. I had to organize my own research group and help design our lab facilities in Chernogolovka. We also had to develop new directions of scientific research. I managed to defend my PhD somewhere between 1956 and 1960. I don't remember the exact date. What I do remember, though, is Semenov's visit to Chernogolovka on September 24, 1960. He came to Chernogolovka to discuss new directions of scientific research with us. I remember it so well, because it was my birthday and after we finished work, we made shashlyk[7] and had a little celebration right outside the hotel. We also began recruiting young employees as early as 1956. I was only twenty-six years old, but I was already recruiting college seniors to work under my direction.

Did you know back in 1956 that you would have your own lab?

No. You see, we didn't worry too much about the administrative side of things, and we certainly didn't feel entitled to having our own labs. Our goal was to create new directions of scientific research, and we concentrated all our efforts on that. By the time I was put in charge of my own lab, I already had fifteen people working under my direction. Most of us came to Chernogolovka almost every day, since we also had to supervise the construction of new lab facilities. We stayed at a small hotel on First Street: Dubovitskii, me, and three of my colleagues, who would also become the first lab directors in Chernogolovka. Gradually, as the construction of the first housing and research facilities went forward, more and more people moved here permanently.

Georgii Borisovich, you mentioned earlier that Semenov envisioned building not just a testing ground, but a scientific center in Chernogolovka, the so-called "Russian Cambridge." Can you tell me a bit more about his vision?

From the beginning, Semenov planned to expand the scientific research carried out at the testing ground. This is why he put me and other young scientists in charge, asking us to develop new directions of scientific research. Semenov's main goal was to advance fundamental research, and for him the testing ground was a means to that end.

What surprises me is that, while the Communist Party was heavily involved in the construction of Akademgorodok or Pushchino, this was barely the case in Chernogolovka.

You know, this can be attributed to Semenov's peculiar style of operating. There were two ways to get things done in the Soviet Union. First, one could go to the very top of the party hierarchy and lobby there, usually overcoming a lot of resistance. In this scenario, the state had to allocate large sums of money all at once. The second way was to simply start the fight, so to speak, hoping that things would work out at the end. Semenov preferred the latter. Even though he envisioned building a scientific town in Chernogolovka, he didn't share this vision with anyone. He simply told the Soviet government that he would build a high-quality testing ground. Only gradually, in the early 1960s, did he start transforming the testing ground into a scientific town, creating conditions for the establishment of new research institutes here. He thought that if theoretical and experimental physicists, chemists, and biologists all lived and worked in the same place, they would communicate with each other, even outside of work. He hoped that this would lead to the exchange of ideas and eventually scientific breakthroughs at the intersections of different disciplines. This interdisciplinary dimension was extremely important to him.

Georgii Borisovich, what did everyday life in Chernogolovka look like?

Well, the material conditions of scientists kept improving beginning in the late 1940s, when the Soviet government realized that science would play a crucial role in the economic and military development of the country. Scientists received some of the highest salaries in the USSR. For example, the salary of a head of a lab, who had a doctor of sciences degree, was 500 rubles. This was as much as the salary of a Soviet minister. Of course, ministers also enjoyed various other benefits, including free meals, access to free cars, and paid vacation. But to give you a point of reference, in the early 1960s a junior researcher in Chernogolovka, who had just graduated from college, received a salary of 105 rubles. This was considered to be a good salary, since 150–200 rubles was considered to be a truly substantial salary across the Soviet Union.

Besides, the state allocated free housing to scientists. In Moscow, and in other parts of the Soviet Union, one would wait for years to receive housing. This was not the case in Chernogolovka, where housing construction was booming. My family, for example, got an apartment shortly after we moved here, and we paid next to nothing for communal services. You see, at the time scientists were considered to be the elite of Soviet society. Even though their material conditions were not as good as the material conditions in the West, they were still much better off than the rest of the Soviet population. For the first decade or so, nearly 100 percent of newly arriving employees received housing right away, which was extraordinary for the Soviet Union.

One other thing I want to mention is that since Chernogolovka was built from scratch, we had a lot of say about what the town would look like. We wanted to live in a green community, so Chernogolovka was built with lots of parks. We planted some trees ourselves, since we felt like it was our town, and we were in charge. We usually did this type of work over the weekends. When foreigners were finally allowed to come visit our town in the late 1980s and 1990s, they really liked it here.

At first, though, living in Chernogolovka was a real challenge. In the early 1960s, there were no grocery stores here. There was no direct road connecting Chernogolovka to Moscow either. In 1964, I became one of the first few residents in Chernogolovka who bought a car. I remember it cost me 6000 rubles. It was a gray "Volga." Of course, it was not that easy to buy a car in the Soviet Union. Each organization only received a limited number of cars, which it then distributed among its employees. Say the Academy of Sciences would get 100 individual cars. We then had to sign up and wait in line, sometimes for several years, for our turn to buy a car.

Georgii Borisovich, did you participate in organizing cultural events in Chernogolovka?

Well yes, I did. In the mid-late 1960s, several scientists, including Lev Gor'kov, Vladimir Broude, Lev Atovmian, and myself, formed the so-called initiative group. We would meet on a regular basis, mostly in my apartment. Our goal was to improve cultural life in Chernogolovka. Dubovitskii was already planning to build a cinema theater here, and the Academy of Sciences allocated funding for its construction. But we wanted to expand the cinema and transform it into a cultural center, the so-called House of Scientists. We hoped this would allow us to invite artists and lecturers from Moscow. And most importantly, we wanted to create an informal cinema club in Chernogolovka. It was called "Kaleidoscope." So Dubovitskii made the construction of the House of Scientists possible, but the rest was up to us.

Why did we need an informal cinema club, you might wonder? It allowed us to see movies that were inaccessible in regular cinema theaters. We had access to a special film repository near Moscow, and borrowed foreign films from there, as well as Soviet films that had been "shelved" so to speak. "Kaleidoscope" also allowed us to invite film directors who were not part of the mainstream Soviet film industry, like Andrei Tarkovskii. The fact that Vladimir Broude, a member of the Communist Party and a prominent scientist, was formally a head of the club made it much easier for us to borrow all these movies.

And who sponsored this informal cinema club?

No one did. The cinema club was a part of the House of Scientists. Besides, we didn't need much money to run it. The building itself belonged to the Academy of Sciences, and they didn't charge us anything. We also introduced small membership fees. Besides, back then artists didn't charge much for their concerts: 100 rubles was considered a good honorarium. Sometimes we brought films to Chernogolovka even before they reached the broader public. Tarkovskii, for example, came here to present his film, *Andrei Rublev*, before it was even shown on the big screen.[8]

Did you ever get in trouble with the regional party committee?

No, not really. Chernogolovka's party organization was part of the October regional committee of the Communist Party, whose employees were much more open-minded. I already mentioned that we had Broude, who was a party member, as a head of the club. Besides, we didn't advertise any of these events outside of Chernogolovka. We organized them for ourselves. You also need to understand that towns like Chernogolovka enjoyed a special status. Most people who lived here were members of the scientific intelligentsia. There was much more intellectual diversity here than in the rest of the Soviet Union. Since we worked for the state and fulfilled various state assignments, party officials trusted us more. They knew that there were no openly anti-Soviet views among local scientists. I would say that both the party and the KGB tolerated the diversity of viewpoints among the scientific intelligentsia because they knew that, ultimately, we worked for the state.

You see, back then scientists were considered to be among the elite of Soviet society. Soviet rulers had a lot of respect for scientists, since they believed that they needed science to build Communism. Both Lenin and Stalin wrote scientific books and articles. It is true that Stalin did a lot of harm, too, especially with his attack on biology and his support of Trofim Lysenko.[9] But overall Stalin thought that science was indispensable for Soviet economic and military success. Soviet rulers recognized that the country could not advance without science, and they invested a lot of money in it, including building numerous research institutes and entire scientific towns from scratch. By the way, some of the most prominent Russian research institutes were established in the 1920s, when Russia was still recovering from the disastrous Civil War.

Besides, when we came into the scientific profession, it was a completely different era. It was certainly different from 1937, when scientists could be arrested and shot without any explanation. It was not 1948 either, when Stalin was actively interfering in various scientific disciplines. The relationship between scientists and the state in the mid-late 1950s was not as politicized as it used to be. There was a genuine interest—on the part of Soviet officials—in what scientists had to say. A lot of Soviet ministers and top party officials visited Chernogolovka to see what kind of research we conducted here. Many Soviet ministers, by the way, had some sort of technical background, which allowed them to better understand the problems at hand. Soviet authorities were also interested in getting good results, and they encouraged us to argue with them. They didn't have to worry about staying in power, and they were more invested in getting things done.

Of course, we didn't like certain things that the Soviet government did. But at the end of the day, I would agree with Valery Legasov, a chemist and a member of the Soviet Academy of Sciences, who worked at the Kurchatov Institute. He said that our highly centralized system was well designed to solve large scientific and military problems, such as building nuclear plants or missiles. But it was ill-suited to tackle smaller tasks, such as organizing food supplies. You should also keep in mind that most people who came to live and work in Chernogolovka were physicists and chemists. Soviet authorities had to listen to us, while they could be more dismissive of scholars in the humanities. The highly centralized Soviet system presented some problems to us as well, but they were much easier to solve, since we got a lot of money from the state. Moreover, Soviet authorities tolerated more diversity of thinking among the scientific intelligentsia. Party officials recognized our shortcomings, but they knew that there were no anti-Soviet people living here.

But was there some kind of unspoken boundary that scientists could not cross without getting in trouble?

There was, but it was extremely subtle. Overall, scientists' lives were not as politicized as the lives of historians or philosophers, for example, where the party imposed its guidelines on the entire discipline. Even Stalin hesitated to interfere in physics. He tried once, but the physics community told him that he had to choose between ideological purity and the nuclear program. He chose the latter, as you know. But political questions were never a priority for scientists. They cared much more about their work and their immediate surroundings than politics. In fact, most people care much more about their everyday lives, family, and friends, than some bigger problems. This is natural.

Were there any people among your friends or acquaintances who experienced repressions first hand?

Of course, there were. I already told you about my uncle. But mass repressions ended at some point, and we found out about them only years later. My cousin, for example, managed to rise up the party ladder and became the second secretary of the Central Committee in Kazakhstan, despite his father's arrest. My good friend and colleague, Vadim Barzykin, also lost his parents in the repressions. For some people, this stigma was deadly, but others managed to overcome it. Later, in the 1960s and 1970s, Soviet officials stopped asking about this in the official questionnaires. While in the 1930s and early 1940s children of "enemies of the people" were not allowed to apply to a university, this changed after the war.

Do you think that people, whose families had been unfairly persecuted, experienced some kind of internal discomfort or tension which impacted their views on the Soviet regime?

They probably did, but it was hard to measure. Even *One Day in the Life of Ivan Denisovich* was not all that black and white. I read it back in the 1960s, when it was first published in *Novyi Mir*. Overall, though, the Soviet intelligentsia viewed many slogans of Soviet power positively. It was not by accident that many members of the intelligentsia supported the Russian Revolution. Even though the practical applications of these slogans turned out to be problematic, the intelligentsia could relate to many of them. Besides, the Russian Civil War polarized Soviet society. There was a lot of violence during the war on both sides.

You should also keep in mind that many members of the Soviet intelligentsia grew up under Soviet power, and owed their success to the state. Many shared the views of Soviet authorities. It was also extremely important that scientists were in demand in the Soviet Union. I already mentioned the Leningrad Physico-Technical Institute, which was established during the Civil War. But there were plenty other research institutes that were founded in the 1920s and 1930s, including the Institute of Chemical Physics, where I started working, which was established in 1931. Scientists recognized the shortcomings of Soviet power, but overall, they shared the interests of the state. They put up with many things they didn't like in the name of these shared state interests. Life is complicated: it is never just black and white, there is a lot of gray in it, too.

Marx, by the way, was a smart person. He has become popular in the West again recently, after the economic crisis. Marx wrote about class conflict within one country, but now his theory has become relevant for the entire world. The economic exploitation that Marx described is now taking place among countries. It is well known, for example, that the United States and Europe consume a large portion of the world's resources. As the US national debt keeps growing, this exploitation will become more severe. But at the end of the day, other countries will not want to tolerate the fact that 40 percent of world resources are consumed by one country. So the class struggle that Marx described in the nineteenth century has evolved and taken the shape of globalization.

It sounds like you're saying that there are problems with both capitalism and socialism.

Well, I can relate much better to true socialism, not the socialism-of-1937 variety. Nowadays, you can sometimes hear people compare Communism and Nazism, arguing that they were not that different. But there was a big difference between them. One ideology was fascism. It promoted nationalist agenda and justified the struggle for the living space, as well as the destruction of the Slavs, Jews, and other nationalities. Our official ideology was quite different. Of course, Communism was a utopia, because it didn't take into account the biological needs of human beings. Still many people could sympathize with its main slogan: "From each according to his ability, to each according to his needs." I think Sweden has made the most progress in this direction. China has also been quite successful. You know Deng Xiaoping was an extremely intelligent ruler, unlike Gorbachev. Most Chinese are content with their lives today. Of course, when I visited China, I met mostly scientists. But China has experienced tremendous economic growth over the last several decades, which improved the material well-being of its people.

Are you saying that the socialist experiment in China was successful?

They carried it out differently. Besides, what they have in China now is not really socialism. But they have taken steps in the right direction. Back in the 1970s, when they started reforms, they didn't begin with glasnost or political freedoms, like we did. Instead, the Chinese focused on improving the economic situation and raising the living standards. They allowed some private property, which led to some people getting wealthy. The state still owns most resources and corporations in China, and it controls them with an iron hand. This has facilitated economic growth. The Chinese never paid too much attention to political freedoms, although we can see some progress in this direction as well. And overall their development has been peaceful. True, they have extremely wealthy people, too, but the transition in China was not as drastic as in post-Soviet Russia. Moreover, today, China can compete with the United States in the economic sphere. Of course, the question is whether they can transition to a more democratic regime.

Do you think China is moving toward more democracy?

Of course, willy-nilly. But the important thing is that they didn't start with political reforms, as we did, which led to such disastrous results.

I guess you were not a supporter of Gorbachev's reforms then?

I think Gorbachev is a tragic figure. He had all the right ideas: he recognized the need for reforms, including the need for democratization and liberalization. But he was not a talented ruler, and he didn't understand how to carry out these reforms. He started with the wrong things, and this led to terrible consequences.

Do you mean the collapse of the Soviet Union?

Not just that. Gorbachev's reforms led to terrible tragedy of millions of people, whose standard of living fell drastically. They led to the rapid decrease in the length and quality of life of ordinary people. Gorbachev complicated everything. He is a tragic figure, because he had good intentions, but as we know, the road to hell is paved with good intentions. Gorbachev was completely incompetent, and he started his reforms on the wrong foot. I remember listening to one of his public speeches, where he promised that the USSR would become the leader in building automobiles. What an idiotic thing to say!

Now living standards began falling under Gorbachev, but the economic situation grew especially dire in early 1992. After they disbanded the Soviet Union, we witnessed a true economic catastrophe. The reforms carried out by Yegor Gaidar[10] led to horrendous inflation and a sharp decrease in the quality of life.

But didn't Gaidar's reforms help Russia transition to a market economy?

Who cares about a market economy, if one could not afford buying any groceries in the store. I remember we received a large package from our Italian colleagues. When we opened it, we found cheese, milk, and sausages inside. Later, they told us that they saw empty shelves in Soviet grocery stores on TV and thought that we were starving here, so they sent us the package.

Georgii Borisovich, you said earlier that Gorbachev had the right ideas and that many people recognized that reforms were necessary. What did you have in mind?

Well, we realized, of course, that our economy was not developing as fast as it should have. In some industries, we were at the cutting edge, but in others we lagged behind. Our productivity was low. Overall, the Brezhnev-era economy was stagnating, and stagnation is always bad, because there is no economic growth. So many people recognized that reforms were necessary. But we could not have imagined, even in our worst nightmares, what would happen in the 1990s.

It sounds like you are saying that the Soviet Union needed mostly economic reforms?

Yes, but also political reforms. We had formal elections, but we could only choose from one candidate. We knew nothing about this candidate, yet he or she would get elected anyways. The Soviet military campaign in Afghanistan was also a big mistake. Why did we have to interfere in Afghan affairs the first place? And once we decided to leave, why did we abandon our allies there? Everyone knew that the Afghan campaign was a mistake, which had bloody consequences. Everyone knew that our soldiers were dying there. But what happened after the Soviet collapse, when Russia was looted by the oligarchs, was even worse. I detested all these so-called liberals, a lot of whom pursued their own selfish interests. They foolishly tried to copy and recreate the American way of life here. Many of them became millionaires as a result. In the 1990s, they presented themselves as dissidents, but they had nothing to do with real dissidents. In the 1970s and 1980s, these people worked quietly within the Soviet system. In fact, many worked in the Central Committee of the Communist Party and did well for themselves, because they all toed the party line. Most of those people were opportunists and quasiliberals: not only did they work within the system, but they took an active role in shaping Soviet ideology, which they later so eagerly denounced.

The true dissidents of the 1960s and 1970s were completely different. None of them ended up in positions of power after the Soviet collapse. They were idealists, as it often happens. Overall, the scientific intelligentsia viewed these true dissidents positively. Scientists agreed with many things that dissidents were saying, but at the same time, we thought that their open dissent could not bear any fruit. In 1968, for example, several people protested the Soviet invasion of Czechoslovakia on Red Square.[11] They were immediately arrested. Some ended up in jail, while others were released. We could sympathize with these people, but we didn't think that the path they chose was for us. Later, there were also people like Andrei Sakharov, who genuinely fought for his beliefs.[12] He was a brilliant physicist, and I think his dissident activity was a waste of his talents. Sakharov thought that since he was a first-rate physicist, his other ideas were also right. I think his criticism of the Soviet war in Afghanistan was correct, but his ideas for reforming Soviet society were misguided.

In the late 1960s and 1970s, there was, in fact, a dissident who lived in Chernogolovka: Kronid Liubarskii.

I didn't know Liubarskii personally. I only heard about him when he was arrested. After his arrest, I started asking questions and heard mostly positive things about him. Still Liubarskii was an outsider. He was not really related to the Academy of Sciences or the research institutes in the town. He had some friends and colleagues here, but he didn't contribute to the development of Chernogolovka.

Liubarskii and the rest of the dissidents were all idealists. I didn't always agree with their methods, but I could sympathize with their idealism. Besides, most of the dissidents' criticism of the Soviet regime was fair.

Georgii Borisovich, when did you travel abroad for the first time?

In 1964.

Where did you go?

I went to the University of Cambridge. There was a symposium on combustion there, which was attended by many international scientists who worked in the field of combustion and detonation.

I am a bit surprised to hear that your first trip abroad was to England, a Western European country. Was it difficult to go to Britain in 1964?

Well, we had to go through the standard procedure. First, the Academy of Sciences had to include our trip in their annual plan and allocate money for it. Since we traveled to the symposium as scientific tourists, we had to cover part of the travel cost ourselves. Then we had to get the party's approval. We had to submit paperwork to our institute's party organization, which passed it on to the regional committee of the Communist Party of the Soviet Union (CPSU). I am not sure if the Central Committee got involved, but I suspect that our travel also had to be approved by the KGB. A lot of our research was classified, which made the process more complicated. KGB officials basically had to vouch for us that we would not give away any secrets and that we would not try to defect.

What did you think about this procedure?

It was repugnant. Besides, we had to pass an interview with some old party members at the regional party committee. These people usually didn't work anywhere else, but they were ideologically mature, so to speak. They could ask you to name a party secretary of the Central Committee in a country you were going to. Since we were traveling to England, they asked us about members of the British Communist Party. It was ridiculous.

I know that some scientists were not allowed to travel at all. Some were discriminated against because of their Jewish background.

Well, I don't know about that. The Soviet delegation, which I was a part of, had twelve people. Two of them were Jewish. It was a tiny delegation compared to the American and the English ones. I remember Kondratiev, who was one of Semenov's first students and a prominent scientist, was in charge of our group.

And what impression did that trip make on you?

It was incredible. It was my first time abroad, and I went directly to England. And not just England—I went to Cambridge. We lived in an old fourteenth-century building. The fact that we could simply sit and talk on the grass lawn was exciting. The funniest thing, though, was when we discovered that we knew London quite well, even though we had never been there before. The Russian intelligentsia was well read, and we knew many sights, streets, and parks in London from reading Dickens, Wells, and other English writers. When we came to Hyde Park, for example, we knew we should look for Speaker's Corner.

It was also fascinating for us to see everyday life in England, which was so different from our own. I remember we were having breakfast at our hotel one day, and one of the waiters didn't serve us well. "What can you expect," our tour guide said, "He is Irish!" I was shocked to hear this. I could not imagine anyone back home saying "What can you expect from this waiter? He is Ukrainian!" When we went to see the poor districts of London, we also were shocked to see homeless people going thorough trash. Soviet propaganda told us about this, but it was different to see it with our own eyes. There were beggars in the Soviet Union, too, especially after the war, but it was unthinkable to see them going through trash like that, right in the heart of the city.

Were you surprised to see such variety of produce in grocery stores?

Yes, but in the mid-1960s, we had enough food in the Soviet Union, too. We only started experiencing food shortages toward the end of the Brezhnev era and during the Gorbachev era.

So I guess your trip didn't make you wonder why people in England lived so much better that people in the USSR?

No, it didn't. It is true that we didn't have certain things, especially clothes and electronics. But in terms of food, we could buy many things back home.

How often did you travel abroad in the Soviet period?

Well, I went to England in 1964. Then I traveled to Czechoslovakia several years later, then France in 1968. I even went to the United States. I had to go through the procedure I already described every time. But I was never refused permission to travel. This was partially because we conducted research on solid rocket propellant, which was important for the state. Besides, I already went abroad, and it went well. I didn't cause any trouble, so they knew they could trust me. You see, going to conferences was one of the few ways to receive the most up-to-date scientific information, and Soviet party officials recognized that. I think every time a scientist went abroad, they calculated the risks and decided if the benefits outweighed those risks.

And how important were your interactions with foreign colleagues for you personally?

They were extremely important. Science knows no national boundaries.

I am asking because some of your colleagues told me they were not allowed to travel abroad, but they said it didn't impede their work.

Some people didn't want to travel as much, even around the Soviet Union. But my other colleagues and I were curious, and we went to many different countries.

From where were the scientists with whom you usually collaborated? How did you establish the initial relationship?

You see, we worked on combustion, detonation, and rocket propellant. Most countries that could afford investing money in these areas were technologically advanced countries, like the United States, England, France, and later Germany and Japan. We also collaborated with the Czechs and Poles in Eastern Europe. After the reforms, China joined in. We usually met foreign scientists at international conferences and started corresponding with them via mail. Sometimes they invited us to come over and work in their labs.

Did you speak any foreign languages?

Yes, I studied German. I also learned enough English to explain my research, since we used a lot of the same vocabulary. I asked to translate some of the longer papers for me.

Did you ever think about emigration?

No. In the late 1980s, I was invited to come and work in the United States, but I refused. I love Chernogolovka. It is our child. I like Russian culture, food, and our way of life in general. When I travel abroad, I always want to come back, even if I am having a wonderful time.

I know there were some scientists who defected to the West. What did you think about those people?

I personally only knew one scientist who refused to return to the Soviet Union. He was a theoretical physicist and worked at the Institute of Chemical Physics in Moscow. I remember my colleagues and I thought badly of him, but it was a complicated story. We disliked him because when he asked for political refuge he also bad-mouthed a lot of great scientists at ICP, like Khariton and Zel'dovich.[13] There was another scientist at the ICP Branch in Chernogolovka, who claimed that he needed to leave the Soviet Union, because as a Jew he could not advance professionally. This was not true: by that time, he had made a successful career for himself and was head of an entire scientific department. He simply wanted to leave, and we thought it was foolish. We could not understand why people like him wanted to leave. We were all part of one community, where everyone was more or less equal. What was so bad about it?

Well, maybe he wanted to be different.

Maybe. He left and we never heard anything about him. In fact, most people who left had to take worse jobs over there than they had here.

Georgii Borisovich, many members of the nineteenth-century Russian intelligentsia looked to the West and compared Russia to the West. Did the Soviet intelligentsia also do this?

This is a complicated question. As far as science goes, we always thought highly of Western science. But we never thought of Western science as superior. They were better at some things; we were better at others. I would say we felt that we were equal. And I think they agreed with us. Take my first visit to England, for example. The Congress took place at the University of Cambridge. The honorary chairman of the congress was Ronald Norrish, who was a Nobel Laureate and fellow of the English Royal Society. On the second day of the Congress, he invited us to his office. There were about 800 people at the Congress, but he only invited a group of American scientists and us. We talked about the need for closer collaboration between American and Soviet scientists. Eventually, he invited select members of the Russian delegation to his home. Nowadays, the attitude to Russian scientists is very different, which is directly linked to the general decline of science in post-Soviet Russia.

NOTES

1. The Supreme Soviet of the Uzbek SSR held the nominal legislative power in the republic during the Soviet era. The republic was *de facto* ruled by the Communist Party of Uzbekistan, which was a part of the Communist Party of the Soviet Union (CPSU).

2. Valery Legasov was a prominent Soviet chemist who helped investigate the Chernobyl disaster of 1986, but committed suicide two years later.

3. The first Soviet nuclear bomb was tested on August 29, 1949, at the Semipalatinsk test site in present-day Kazakhstan.

4. The Leningrad Physico-Technical Institute was founded in 1918 and played a central role in the development of Soviet physics.

5. The New Economic Policy (NEP) was the economic policy implemented by the Soviet government from 1921 to 1928. It included the abolition of grain requisitioning (which was replaced by a tax), as well as the abolition of the state monopoly of small- and medium-scale manufacture and retail trade.

6. The Ministry of Medium Machine-Building was in charge of the Soviet nuclear industry. Semenov's institute participated in the Soviet nuclear project since 1946.

7. Shashlyk is a popular Russian cookout meal, similar to shish kebab.

8. Even though Andrei Tarkovskii finished *Andrei Rublev* in 1966, the film was shelved for almost five years. The censored version came out in cinema theaters in December 1971.

9. Trofim Lysenko was a Soviet agronomist and biologist, who led an assault on Mendelian genetics during the Stalin era. He rejected Western genetics in favor of a Soviet theory that emphasized the inheritance of acquired characteristics. His pseudoscientific approach caused a lot of harm to Soviet agriculture and biology.

10. Manelis is referring to the radical economic reforms carried out by Yegor Gaidar, the Prime Minister of Russia under Boris Yeltsin, in 1992. Known as "shock therapy," the reforms aimed to transition Russia to a market economy, through nationwide privatization and freeing of prices. Although beneficial in the long run, Gaidar's reforms led to a steep inflation and mass poverty.

11. The Soviet army invaded Czechoslovakia on the night of August 20–21, 1968. The goal of the invasion was to crush the Prague Spring, a series of reforms, which

called for more openness and partial economic decentralization in Czechoslovakia. Several days later, eight Soviet dissidents protested the invasion with a sitting demonstration on Red Square. The protesters were beaten and arrested. Most were sentenced to several years in the Gulag.

12. Andrei Sakharov was a prominent Soviet nuclear physicist and a key participant in the nuclear project. He became a leading Soviet dissident later in life.

13. Yulii Khariton and Yakov Zel'dovich were prominent Soviet scientists and leading participants in the Soviet nuclear program. Both were students of Nikolai Semenov.

Chapter 4

"I Have Always Embraced Collectivism and Had Little Tolerance for Individualism"

Lev Nikolaevich Vashin

Lev Nikolaevich Vashin (1929–2015) was one of the few Chernogolovka scientists who maintained his faith in the Communist ideology even after the Soviet collapse. Unlike most of his peers who willingly gave up their party membership in the last days of the Soviet empire, Vashin never disaffiliated from the Communist Party. In his interview, he described a brief period when the Communist Party was banned in post-Soviet Russia (1991–1993) as a "forced interruption." He restored his party membership in 1993, as soon as the Communist Party of the Russian Federation was established. Although critical of Stalin's mass repressions and "his anti-party activity," in the interview Vashin praised Stalin's leadership during the Great Patriotic War, as well as the increased funding of Soviet science in the late 1940s. In his early eighties, my interviewee remained a staunch believer in Communism, convinced that it better suited the needs of ordinary people. Neither the truth about the Communist atrocities in the Soviet era, nor the collapse of the USSR in 1991 undermined Vashin's faith in the Communist cause. As I conducted two interviews with this accomplished scientist in March and April 2010, I tried to identify the origins of such unwavering commitment.

Lev Nikolaevich's family background and early upbringing shed some light on this question. Both of his parents came from impoverished peasant families and benefited tremendously from the October 1917 Revolution. Vashin's father, a son of a village shoemaker, left the countryside shortly after the Civil War. He worked briefly as a stoker at a Komsomol committee in a small town in Belarus. Later he attended several courses to advance his education, which made him eligible for teaching positions during the nationwide campaign against illiteracy in the 1920s. He continued to climb the social ladder, joining the Communist Party in 1927 and participating in the collectivization of his own village in 1929. By the early 1930s, Vashin's father became part of the emerging Soviet "middle class," serving as a director of a middle school, and later a trade school. The family moved to Kashira, a town seventy miles south of Moscow, shortly before the Great Patriotic War. Lev Nikolaevich fondly remembered growing up in a communal apartment in Kashira—one of the nineteen apartments in a building populated by sixty-four families. In the interview, he argued that all residents of the building "lived in harmony" and that, since most residents came from the countryside, the Kashira authorities did their best to accommodate their old habits. It appears that Vashin's attachment to collectivism and his rejection of individualism had their roots in his early upbringing.

Vashin's social background made it easy for him to enroll in the elite Physical-Technical Department of Moscow State University (MGU) in 1946, since the Soviet state gave priority to students of working class and peasant origins. Lev Nikolaevich graduated from Moscow University in 1952 and was immediately able to secure a research position at the Institute of Chemical Physics (ICP), directed by the Nobel Prize laureate Nikolai Semenov. The growing demand for scientific specialists during the early Cold War era facilitated Vashin's remarkable professional success. At the age of thirty-one, he was put in charge of his own scientific lab at the Branch of the ICP in Chernogolovka. This impressive promotion made him part of the select group of scientists living in the already privileged scientific community. Upon their arrival in Chernogolovka, for example, Vashin and his family received a separate two-storied cottage, which was an unattainable dream for the vast majority of Soviet citizens. Ironically, in his interview Vashin criticizes the social stratification in post-Soviet Russia, seemingly unaware of the fact that he himself was a member of the materially privileged Soviet elite.

One of the most noticeable features of Lev Vashin's interview is his reluctance to discuss the repressive side of the Soviet system. Following his father's example, he applied for a party membership in 1954. He continued to view himself as "a loyal soldier of the party" even after Khrushchev made his shocking revelations at the Twentieth Party Congress in February 1956. Throughout the interview, Vashin repeatedly defended the legitimacy of a one-party state and rejected the notion that the USSR was a dictatorship, emphasizing instead the popular support of Soviet power. Vashin explained that the Communist one-party state aspired to serve the Soviet people. He admitted that it failed sometimes, but it still better suited people's collective needs. Even more strikingly, Vashin defended the Soviet invasion of Czechoslovakia in 1968 and justified state repressions against Soviet dissidents. Both views became widely unpopular among the Russian intelligentsia in the post-Soviet era. Claiming that he never felt restricted in what he could say or read, Lev Nikolaevich criticized the dissident struggle for the free flow of information, which he thought was overall harmful to Soviet society.

Vashin's enduring commitment to the Communist cause demonstrates the extent to which some Soviet citizens internalized the official ideology and propaganda. His reluctance to give up his views even decades after the Soviet collapse points to the challenges of Russian society's transition from authoritarianism to democracy in the post-Soviet era.

I was born a long time ago, in 1929, in Belarus. Both of my parents came from a peasant background. My mom only had a basic education: she finished parochial school. My dad was somewhat luckier. Since he was an exemplary student, they sent him to continue education at a city school, where he studied for two more years. After the Russian Civil War ended, my dad decided that he was more educated than the rest of the young people in his village. So he left the countryside and moved to a nearby town, where he found a job as a stoker at a Komsomol committee. He was noticed almost immediately there. They sent him to take several courses, which he completed. So even though my father never received a formal education, he was relatively well educated for his time. He served in the army for a while, and was discharged. Later they launched a nationwide campaign against illiteracy. But since they didn't have enough teachers, they recruited almost everyone who knew slightly more than an average person. This is how my father became a teacher. By 1932, he was already a director of a middle school in a small railroad community in Belarus, not far from the old Polish border. Imagine, they could not find anyone better than him for this position! Later my father was drafted again. They sent him to Moscow for rigorous military training. He studied in Moscow for one and a half years until he got sick and was discharged again.

My dad started looking for a job in Moscow and was eventually offered a job as a director of *rabfak*.[1] This was a kind of evening school for workers and peasants. While he was there, my mom also studied at *rabfak*. My dad didn't stay there long, though. During the Soviet-Finnish War,[2] he was drafted again, then demobilized. He worked as a head of a trade school for a while. After the Great Patriotic War broke out, we evacuated to Altai, where my dad was put in charge of a middle school again. Later he was drafted and served at the frontlines. But his health was bad, and they eventually let him go. We went back to the Moscow region. My dad worked as a director of a trade school for some time, but he didn't like that job. Later he switched to a career in administration, serving as head of a personnel department at a factory. My mom didn't work. She was a housewife, like most women at the time. Only during the war, when my dad served in the army, did she take a job as a nighttime guard at a school.

What subjects did you father teach at school?

He was a teacher of history. After we moved to Altai, we settled in the city of Oyrot-Tura, present-day Gorno-Altaysk. My dad was a director of school there. When I finished sixth grade and started seventh grade, I took a class with him on the Constitution of the USSR.

Was it a class on the Soviet Constitution of 1936?

Yes, of course. Everyone had to study it. I went to school in 1936, at the age of seven. I finished the first five grades before the war, at a school in Kashira. At the time, the town was called Kaganovich; only later was it renamed Kashira. The town where I grew up was famous for the Kashira Power Plant, which was built as part of the GOELRO plan.[3] Then the war began and we had to evacuate. We returned from evacuation only in the fall of 1944. But we could not go back to Kashira, since our room had been occupied by other people during the war. It was quite common at the time, even though many of our belongings were still there. My mom's brother lived not far from Moscow, in Lyubertsy. He had a small house there, plus he had an apartment in Moscow. We moved into that house, so I finished middle school in Lyubertsy.

After graduation, my friend and I decided to apply to Moscow University's Department of Physics. Originally, the competition there was not bad at all: two people for one spot. But after the war ended, all veterans were admitted to university without any exams. This completely undermined my calculus: now the competition became ten people for one spot. As a result, my friend and I didn't get in. Since we didn't want to lose a year, we applied to study at the Department of Chemistry at the Moscow Institute of Fine Chemical Technology and were successfully admitted.

Why did you decide to apply to the Department of Physics in the first place?

You see, growing up I was not passionate about anything in particular, which was probably not a good thing. I didn't have much interest in technology. Neither did I have any talent for the humanities. So I decided that the best thing for me to do was to study hard sciences. Since I didn't get into the Department of Physics on the first try, I began studying chemistry. I had been enrolled at the Institute of Fine Chemical Technology for about a year, when a new Physical-Technical Department opened at Moscow University. In 1946, the government issued an order to admit senior students from other higher education institutions to this department. My friend and I immediately decided to transfer. We had to pass two additional exams: math and physics. Since I had been studying chemistry for a year, I applied to study chemical physics. Nikolai Semenov, himself a member of the Soviet Academy of Sciences, was present at our entrance exams. This is how in 1946 I became a student of MGU.

It was an unusual department, I should say. Most of our teachers were members of the Academy of Sciences. There were a lot of prominent scientists among them, including Kapitsa, Semenov, Keldysh, Landau, Tamm, Khristianovich, and Alikhanov. Some of them were directly involved in the nuclear project. Students could major in nuclear physics, chemical physics, radio technology, optics, and aerodynamics. The main goal of our department was to produce scientists for research institutes. You see before the war most technical colleges educated engineers to work in industry, while universities prepared college teachers. Yet few universities produced specialists for scientific research institutes. Around 1940–1941 a group of prominent scientists wrote a letter to the Soviet government, in which they asked to launch a program that would train such personnel. But the war got in the way. Only after the Americans tested their first nuclear bomb in July 1945 did Stalin recognize the importance of this project. He immediately ordered to set up various laboratories and design bureaus. Igor Kurchatov was put in charge of the nuclear program, since before the war he had directed a laboratory that conducted nuclear research at the Leningrad Physico-Technical Institute. The establishment of the Physical-Technical Department at MGU was part of this large-scale effort. It was designed to prepare specialists for the nuclear project. In fact, more than half of my classmates were assigned a job in Arzamas-16, present-day Sarov.[4] Those who specialized in aerodynamics were sent to TsAGI in Zhukovsky.[5] The Institute of Chemical Physics was not directly involved in the nuclear project, but in 1953 it received a state assignment to conduct research on new powerful explosives.

The only problem was that in 1951 they decided to shut down our department and transform it into a separate institute: the Physical-Technical Institute. It presented certain challenges for us. I remember I defended my thesis on December 29, 1951, and had to take final exams in early January 1952. On January 16, 1952, I was already employed at the Institute of Chemical Physics. However, we found out that we would not get an MGU diploma. We went to the university administration to protest. We said that we had studied at Moscow University and wanted to have university diplomas. Eventually, they acquiesced. So I am one of the few graduates of Moscow University who has a diploma of the Physical-Technical Department. Afterward, all graduates received a diploma of the Physical-Technical Institute. Around the same time, higher education administrators realized that they should prepare scientists for research institutes, not just as college teachers. So from 1945 onward, there was an overall shift in higher education toward greater research support. This was accompanied by the founding of myriad new research institutes and laboratories across the country.

How did this shift compare to what happened in the 1930s?

Let's compare it with the 1920s instead, because already in the 1920s Lenin and the Soviet government recognized the importance of developing science. In 1922, they established the State Radium Institute, which started conducting research on radioactivity. I am not sure if anyone conducted such research in pre-Revolutionary Russia, but I know that as soon as the Civil War ended, radioactivity became the focus of research at the Radium Institute and the Leningrad Physico-Technical Institute.

Lev Nikolaevich, you mentioned that both of your parents came from peasant backgrounds. What did they think about the October 1917 Revolution?

They viewed it very positively. Why, you might ask? They both came from impoverished families. My father came from a family of six. His grandfather was one of the three brothers, and before he died he divided his land among his sons. My father's father received a tiny plot of land. To survive, he became a shoemaker. So my family was making shoes for an entire village. They had some land, of course, and they plowed and sowed it, but they also made shoes for village residents. But as I already told you, my father left his village shortly after the Civil War. He worked at a local Komsomol committee. And in 1927 he joined the ranks of the party, the VKP(b).[6] My mother's family had six children, too. Her father inherited no land at all. He was a hereditary laborer, and for the last eighteen years of his life, he worked at a mill. These are my ancestors. I already mentioned to you that my father was drafted into the army numerous times. In the 1930s, they even sent him to the Far East. There was a conflict over the Chinese Eastern Railway there, but it had ended by the time he arrived. When he returned, they sent him to his native village to carry out the collectivization campaign[7] there.

Did they want your father to collectivize his own village?

Yes, in 1929. By the way, my grandfather was one of the first people to join the kolkhoz.[8]

How did the collectivization campaign proceed in your father's village?

It was peaceful. There were no incidents with the kulaks. They simply combined all the land. My grandfather's plot was only 2 hectares, even though an average peasant plot of land in Russia was 10 hectares at the time. So he benefited from collectivization.

Were there any well-off peasants, or the so-called kulaks, in the village? Was anyone de-kulakized?

I don't know anything about that. My father didn't tell me. But you see a Belorussian village was different from a Ukrainian village. Ukraine had a much better climate and richer soil, which means they had much better harvests. Ukrainians were better off as a result. Belorussian peasants were poor on average, and they didn't object to collectivization. The villages where my parents came from were not far from each other. They both belonged to Count Chapsky. I saw this count's estate in 1938. The Bolsheviks organized an orphanage there, while they transferred all his land to the peasants. But the land was barren.

Do you remember the atmosphere in the Soviet Union in the 1930s?

I became self-aware around the age of seven. We lived in a large building that had many communal apartments. Most residents had come from the countryside and worked on building the Kashira Power Plant. There were several families of the intelligentsia living there, too. Our building had nineteen two- and three-room apartments. Each room was occupied by one family; sometimes two families shared a room. Overall, about sixty-four families lived in our building. Since most of them had come from the countryside, they still had many old habits, which city authorities tried to accommodate. For example, the authorities built barns on the outskirts of Kashira, where local residents had cows, goats, and pigs. One could see cows or goats grazing nearby. This was the atmosphere of my childhood. Nowadays, everyone criticizes communal apartments. What can I say to this? There were certain problems with communal living, of course, but overall people lived in harmony, to say nothing of us children. There were about fifty children of different ages living in our building, and we were all part of one big group. Our building had central heating and electricity, although there was no hot water. Every apartment had a kitchen with a wood-burning stove. There was a barn next to our building, where families kept their firewood. Some people chose to keep chickens there instead. Local children oftentimes took initiative, too. For example, Natasha Kuleshova organized amateur theatrical performances for children, which usually took place in a barn.

Where did most residents of Kashira work?

The majority of people worked at the power plant. Women were mostly housewives, few of them worked. There were various stores, a dressmaking establishment, a barber shop, and a canteen in our town.

Did the mass repressions of the 1930s have an impact your family?

We heard nothing about the repressions until 1939. In 1939, the father of two of my classmates was arrested.

Where did he work?

He worked as an engineer at the power plant.

What were the charges against him?

I could not know this at the time.

Did you parents talk about it at home?

My parents never said anything about it in my presence. What else do I remember? Our town had a lively cultural life: there was a club, where we had theatrical performances, a cinema, a library, a café, and a radio station.

In 1937, they built a radio network in Kashira. Everyone bought radio sets. I remember my family gathering around a radio set every night to listen to the evening programs.

How big was your family?

My sister and I, our parents and grandmother. My grandmother, by the way, was born into serfdom. She was born in 1860, one year before the abolition of serfdom. She died in 1952. This was my mother's mother. My father's mother died in 1929, my mother's father died in 1918, and my father's father died during wartime occupation, in 1942.

Did you mother take care of your upbringing?

Yes, she did. I rarely spent time with my father. I remember several times he got sick and had to stay at home, so we spent some time together. I also remember that once I decided to slack and didn't do my homework, and all of a sudden my father asked me: "Did you do all your homework?" I didn't dare lie to him and admitted that I hadn't solved my math problems. Even though it was almost my bedtime, my father didn't let me go to bed until I did my math assignment.

Did he want you to be an educated person?

Absolutely. From third grade onward, I was an exemplary student. Of course, it all went awry after the war broke out. I remember one time I was coming back from school and stopped at a bookstore, where I saw a beautiful book, *The Wizard of the Emerald City*.[9] When my dad came home from Moscow later that day, I ran up to him saying: "Daddy, daddy I saw such a beautiful book in the store today." And he opened his briefcase and took out that book. It turned out he already had bought it for me in Moscow (laughs).

Were good books hard to come by at the time?

You know, my parents usually bought paperback editions for me. For example, I had fairy tales by Alexander Pushkin and poems by Samuil Marshak. But we didn't have many serious books at home. To compensate for this, I became a member at my school's library. At first, I only read textbooks, even though I learned to read when I was five years old. But when I was in second grade, a friend of mine showed me a map with Ferdinand Magellan's route and started telling me about this expedition. I asked him where he had learned about all this, and he said he read a book about it. The following day I went to the library and enrolled. I started checking out books on a regular basis. I read them quickly. I also got lucky: when my father became director of a trade school, they started putting together a library. Our neighbor became a librarian there, and she oftentimes brought books from Moscow and kept them in our apartment. While she was entering these books in the catalogue, I could read them. I read many serious books this way, including *The Hunchback of Notre Dame*. In 1941, though, the war broke out and my good fortune came to an end.

Do you remember at what age you became interested in science?

It happened during my first year at the Institute of Fine Chemical Technology. We had a good chemistry teacher there, who organized a student circle, or *kruzhok*. We usually gathered in the evenings to solve challenging problems. We only stayed there for a year, though, until we transferred to MGU.

What was the social background of other students at MGU?

Most students were Muscovites. Some came from the Moscow region. Few people were from other parts of the country.

Were there many students who came from a poor background like yourself?

Well, what can I say to this? When I was a student at the Institute of Fine Chemical Technology, our stipend was 210 rubles. To give you a point of reference, back then 1 kg of rye bread cost 85 kopeks, 1 kg of sugar was 90 kopeks, and butter cost 3 rubles and 60 kopeks per kg. So we could eat fairly well. After we transferred to the Physical-Technical Department at MGU, they increased our stipend to 400 rubles. Now even though we passed all entrance exams in August 1947, our semester at MGU didn't start till November, because we had to obtain a special security clearance.

What did they check?

They checked our family background.

What were they afraid of?

They were afraid of a so-called "bourgeois upbringing."

Did they think that a person with a bourgeois background should not study at the Physical-Technical Department?

Yes, exactly. However, there were few students who came from the intelligentsia milieu.

Did you live in Moscow after you transferred to MGU?

I lived near Lyubertsy during my first year, then moved to Dolgoprudnyi, where the Physical-Technical Department was located.

Do you remember seeing any cases of the anti-cosmopolitan campaign at Moscow University?

No. Although I remember two students were expelled from the university in our second year. They both came from an intelligentsia background. Overall, one could get expelled if it became known that one had relatives abroad.

But people didn't stay in touch with their relatives abroad, did they?

No, they didn't. But sometimes relatives sent packages to the Soviet Union, because they thought people here didn't have enough to eat, and this is how they [the authorities] found out.

Was Stalin's death in 1953 a memorable event for you?

Yes. Our government didn't do a good job organizing Stalin's funeral. I guess they forgot that since Lenin's death the population of Moscow had grown by a couple times. People also came from nearby towns to say goodbye to Stalin. There was a large stampede, as a result. Some people died. As far as our laboratory goes, we discussed what happened, everyone was upset, but no one cried. In fact, when they broadcasted Stalin's funeral on the radio, we gathered around the radio set at work and listened.

Was this your lab at the ICP?

Yes, this was our lab at the ICP.

Did scientists respect Stalin?

I cannot speak for everyone, but among my circle there were scientists who thought highly of Stalin. You see, before 1945 scientists were poor, but in 1945 Stalin issued a decree, according to which scientists' salaries were to be significantly increased. He also introduced longer holidays for scientists with advanced degrees. So Stalin significantly improved material conditions for the scientific intelligentsia. I would even say he overdid it a little bit, because after he died they had to lower some people's salaries. For example, heads of laboratories got 6,000 rubles a month, but after Stalin's death their salaries decreased to 5,000 rubles. All this was before the monetary reform.[11]

Was Stalin's decree related to the beginning of the Cold War?

Well, the Soviet government simply realized that in order to increase industrial and agricultural production, they had to invest more in science and education. As you know, Gagarin's first flight to space made a tremendous impression on the Western world. Western leaders started asking questions, trying to figure out how this could have happened. They concluded that that they needed to invest more in education. I learned about this during perestroika, when we started inviting foreigners to the Soviet Union. I remember there was a symposium on combustion and detonation. It was the tenth symposium, honoring Gagarin's flight. During the reception, an American scientist made a toast. He said that Gagarin's flight made the Americans invest much more in education. But we simply followed Lenin's call to "Learn, learn, learn!" This is how we conquered illiteracy. I already told you about my father, who served as director of a middle school before the war, even though he didn't have any formal education. After the war, however, he realized that his knowledge was no longer sufficient and transferred to an administrative career.

Going back to my childhood, there were about 7,000 people living in Kaganovich at the time. We had an elementary school, a middle school, and an evening school for employees of the power plant. Since there were several villages not far from our town, some village kids studied in my class as well. Overall, we had a good class and a great class instructor. The majority of students were children of engineers, workers, and service personnel.

Did you join the Komsomol?

Yes, I became a member of the Komsomol in 1943, when I turned fourteen years old. Our class instructor gathered all good students in one room and told us it was time to join the Komsomol. I still remember the number of my Komsomol membership card: 19324511.

Quite impressive! Did you also join the ranks of the party later?

I became a member of the party in 1954, when I was already working at the ICP in Moscow.

Did anyone persuade you to join the party?

I had my father's example in front of me. At the time, there was a trend to admit mostly members of the working class, at the expense of the intelligentsia. Since I was a secretary of a Komsomol cell at the university, they admitted me without asking any questions.

As a party member, do you remember listening to Khrushchev's famous "secret speech" in 1956, in which he denounced Stalin's cult of personality?

I guess I am confusing dates a little bit. I was a candidate for party membership, because I was not invited to a party meeting, where they read Khrushchev's report.

Did they read it at a closed party meeting?

Yes.

Then how did you find out about the speech?

My father told me. He came home and said that he heard Khrushchev's speech about Stalin's anti-party activity at a party meeting at work.

And what did your father think about Stalin?

He didn't admire him, but didn't condemn him either.

Was he shocked by the speech?

I don't think it shocked him. But you see, the speech was not just about Stalin. It concerned the entire party, because Stalin was in charge of the party. Of course, my father didn't lose his mind over this. But this was certainly an extraordinary event for him.

What was extraordinary about it?

It turned out that the leader of our party had engaged in the anti-party activity. This was extraordinary.

Does it mean that you first heard about Stalin's repressions from Khrushchev's speech?

No. Actually one of these repressions happened right in front of my eyes. As I already told you, my family left Belorussia in 1934. At first, my father went to Moscow by himself to study, then the rest of us joined him. In 1938, my father got a significant bonus and he decided to take us back to Belorussia to visit our relatives. It was in August of 1938 I think. We visited my dad's father first, then went on to visit my mom's sisters. While we were visiting one of my aunts, an unfortunate incident took place. One morning I woke up to find the entire house (*izba*) full of Red Army soldiers. They came to arrest my aunt's husband. He was charged with spying for Poland.

This is a serious charge.

Yes, it led to his immediate execution. During the war, his son, born in 1919, joined the partisan movement.[12] After the war, he tried to find out what had happened to his father, but to no avail.

And what happened to the rest of the family? Were they repressed as well?

No, but what could they do to them anyways? My aunt worked in kolkhoz all her life. Her son was drafted in 1939. Then the Great Patriotic War began. He was wounded and ended up in a hospital. Then the Germans came and sent all the injured people who were still able to walk to a concentration camp, which they set up 5 km away from my relatives' village. However, the guards there were not good. Some of the prisoners, including my cousin, broke the barbed wire and fled to their native village. Later, as I already mentioned, my cousin joined the partisan movement. He stayed with the partisans until 1944. They blew up German echelons. At some point, Stalin said that if one destroyed a certain number of German echelons, they would receive the "Hero of the Soviet Union" award. My cousin definitely exceeded that number, but he never received an award, because he was "a son of the enemy of the people." I know he was upset about it.

Did your parents discuss the arrest of your uncle?

They talked about it some. My mom was certain that my aunt's husband was innocent. My father was more skeptical. You see, they arrested eight more people from the village that same day. Two of them returned several months later, and they explained to us why my uncle had been arrested. My uncle's village was located 6 km away from the Polish border. This land belonged to the collective farm, and my uncle was put in charge of overseeing a group of workers who were supposed to work on the land. Sometimes he clashed with these workers because they were lazy and he had to force them to work. Eventually, one of them denounced him and he was arrested.

If you knew about the repressions, why were you surprised by Khrushchev's speech?

How could it be different? We were members of the party. The party was supposed to be a leading and organizing force in the Soviet state. And all of a sudden, we found out that there was a traitor in charge of the party. Of course, we were shocked.

What did you think about the fact that there was only one party in the Soviet Union?

We thought it was perfectly normal, at least people in my immediate circle did.

Did you perceive it as a dictatorship of the party?

What does it mean "a dictatorship of the party"? You need to understand that my father was a peasant, and most people in his party cell came from the same background. What kind of dictatorship is it? Did we dictate to ourselves? Rank and file Communists never felt like they were above nonparty members, or that they dictated anything to them. By contrast, party members had to serve as role models for nonparty members.

Do you think people believed that they could build Communism?

When I was little I took it seriously, of course: "From each according to his ability, to each according to his needs." It was not exactly clear to me how we could fulfill this slogan, but we never doubted it. Before I joined the ranks of the party, I didn't really think about how we could achieve that goal. But I thought it was necessary to educate people in a certain way. You see, we were raised to embrace collectivism. Recently, they have been forcing individualism upon us instead. But back in the day we were all together, we were all members of a collective. First, we were Octobrists, then pioneers, then members of the Komsomol. These were all different types of collectives, so we learned to appreciate collectivism from a young age.

Did you identify with collectivism more readily then?

I have always embraced collectivism and had little tolerance for individualism. Since I was little, I have always been an advocate of collectivism and justice. In fact, for me justice and collectivism always went hand in hand. When I was six years old, my mom read Pushkin's fairy tale to me—the one about a fisherman and a gold fish. If you remember, in this fairy tale a fisherman catches a gold fish that can fulfill wishes and he tells his wife about it. His wife, an old poor woman, tells her husband to ask the fish for a new trough. When he brings her a trough, though, she starts scolding her husband, telling him that he had to ask for a cottage house instead. I was little, but it really frustrated me, when I heard this fairy tale for the first time. It was unjust, because the trough was exactly what she had asked for. I am eighty years old now, but I still feel the same way about it.

You see, at some point, there were forty people working in my laboratory. They came from different social and educational backgrounds. Some of them were individualists, others were leaning toward collectivism. But as a lab director, I always tried to organize my lab into a small collective. I wanted everyone to work together. I didn't succeed all the time, but we mostly got along and worked well together.

Lev Nikolaevich, let's go back to the year 1956. You said you were surprised by Khrushchev's revelations at the Twentieth Party Congress. Did this make you question your membership of the Communist Party?

I never questioned my party membership, but I started to notice things that I had not noticed previously. I saw that there were dark pages in the history of the party. But it didn't change my attitude toward the party in general. I thought that as a party member I simply had to keep up with my work.

Do you consider yourself to be a "child of the Twentieth Party Congress," or a member of the shestidesiatniki *generation?*

In 1959 I turned thirty years old, and in the 1960s all my time was dedicated to creating my own lab. It required a lot of work: I had to develop the scientific direction of my lab's research, hire personnel, and divide up responsibilities. All these kept me really busy. Besides, who were these *shestidesiatniki* anyways? They demanded freedom of speech. This was one of their key requirements. But I never felt constrained in this regard. I was always able to say what I thought was necessary. Perhaps, of course, I thought that only certain things had to be said out loud, which is understandable. Once in a while, I even shared political anecdotes with my friends and colleagues. I just didn't like it when people told jokes about Lenin.

Why not?

Because even though I was a materialist, I thought that a man needed to believe in something. Nowadays they say that Communism was not scientific, that it was an ideology similar to religion. Just like religious people believe in the afterlife, Communists believe in building a bright future around the globe. I believed in this, and I still believe that it is possible to build a Communist society somehow. Of course, now I understand that this society would not be the same as the one I imagined back in 1965.

And what kind of society did you envision in 1965?

I imagined that all people would think the same way, and that all people would be inspired by one common idea and would work hard to bring it to life.

Did you think it was feasible?

In the 1960s, I thought it was feasible. Only later did I realize that not all people wanted that. Even so, collectivist thinking was widespread throughout the country. Perhaps 100 percent of the people didn't think this way, but a large majority did.

Was collectivist thinking forced upon people?

It came completely natural to me. My parents were collectivists, too. They never demonstrated any individualist traits.

Would you say that building Communism justified the mass repressions?

No, but I didn't know about the repressions in the 1930s, except for a few separate cases. People started discussing the repressions only after Stalin's death. In this respect, I condemned Stalin, and I don't idealize him today.

I see you have a small portrait of Stalin on your bookshelf.

Yes. I have it here because I am a member of the Communist Party, led by Gennady Zyuganov.

When did you join the party?

I never left the party. The short period when the party was disbanded was a period of forced interruption for me. As soon as it was restored, I renewed my membership.

Does Stalin's portrait symbolize your membership of the Communist Party then?

I never idealized Stalin, but I do think that we could not have won the war without his extreme brutality. He made people work much harder than they would have worked otherwise. I can give you an example from my own life. My family set off for evacuation on foot. On October 16, 1941, there was a panic in Moscow, because the Germans came so close. And on October 18, they forced all vocational schools to evacuate. There was not enough transportation, though, so we started our evacuation on foot. Finally, on December 22, we arrived in Gorno-Altaysk.

Did you walk the entire way? It's very far.

No, we only walked for three days. Then we took a train, a ferry, and a steamboat.

But why do you think the Soviet Union would have lost the war without Stalin?

Because when they evacuated entire factories behind the Urals, these factories were expected to start operating almost immediately. They would build some primitive foundation, install the machinery and some basic electrical equipment, and start working, often under the open sky. Gradually, they would build walls, too. But people worked for 14–16 hours a day in such conditions.

Why do you think this was Stalin's accomplishment?

Stalin set high expectations for people who were in charge. He told them that they needed to produce a certain amount of something by a certain date: fighters, ammunition, what have you. And he used repressions to make people fulfill production goals. If they didn't deliver, he killed them. Even though people didn't have enough food, he made them work at the limit of their physical capacity. You see, Stalin knew how to make people work. If there had not been mass repressions before the war, people would not have worked so hard. But since there were mass repressions, they knew that they had to work hard if they didn't want to be shot. If you ask me about the purpose of Stalin's repressions, by the way, I'd tell you that he simply wanted to create and maintain a myth about his leading role in the Revolution and the Civil War. And he tried to get rid of everyone who knew the truth about it.

Do you think Lenin's cause would have lasted as long as it did if it wasn't for Stalin?

What do you mean by Lenin's cause? You need to keep in mind that the October 1917 Revolution brought a completely new social formation to life. And since this type of social formation had not previously existed, no one really knew what it should look like. There was Marxism, which claimed that money is evil, that it gives birth to capitalists, and that individualism is bad. Later, they said that War Communism[13] was necessary because of the Civil War. But it was really a deliberate policy that the Communist Party tried to implement. They wanted to get rid of money and to introduce barter trade instead: cities would produce industrial goods, the countryside would produce agricultural goods, and then they would exchange. They wanted to abolish money. Later, Lenin realized that peasants could not survive without trading their produce for money, and so he launched the New Economic Policy.

Lev Nikolaevich, let's now talk now about how you ended up in Chernogolovka.

First, I should tell you about the origins of Chernogolovka. You probably already know that the ICP was involved in conducting research on explosives. The institute was originally based in Leningrad. During the war, it relocated to Kazan, and when the war was over the institute's director Semenov decided to relocate the institute to Moscow. I already told you that I studied at the Physical-Technical Department of Moscow University. I majored in chemical physics, and the expectation was that everyone with this major would work at the ICP. This is how I ended up at the ICP. Now, the idea to build a testing ground of the ICP in Chernogolovka belonged to Semenov. He needed a testing ground to carry out research on new powerful explosives. They ran their first tests in Moscow, working in special facilities built of reinforced concrete. But since the institute was located in the center of Moscow, there were significant limitations as to what kind of tests they could carry out there. This is why they decided to establish a testing ground outside of Moscow. Since I wrote my university diploma on explosives, I ended up working at the laboratory that studied explosives at the ICP. This is how I ended up in Chernogolovka.

Were you still a graduate student at the ICP, when you moved here?

No, by the time I transferred to work in Chernogolovka, I had already defended my candidate of physical and mathematical sciences degree.[14]

Did you work at Fedor Dubovitskii's lab at the ICP?

No, I worked at the laboratory of Aleksandr Fedorovich Beliaev. Later I transferred to work under Alfred Ivanovich Apin, who had previously spent some time in Sarov, working on the nuclear bomb.

Was your lab also involved in research on the nuclear bomb?

No, we studied the theory of detonation of explosives, focusing on the speed and the temperature of detonation.

But was the ICP involved in developing the nuclear bomb?

It participated in its testing. Our employees were present at the testing ground. They recorded the speed of the blast wave, gauged radioactive emissions and electromagnetic fluctuations. The ICP designed devices for all these.

Did Semenov intend to build a town in Chernogolovka?

Semenov was an experienced man, and he realized that he should not be too ambitious in his plans. He started off with a modest idea: to build a testing ground with several laboratory buildings and testing sites. As far as housing goes, they planned to build only one residential street. At first, the Academy of Sciences was in charge of construction, but it didn't have sufficient resources. So Dubovitskii appealed to the Council of Ministers, asking to allocate construction to a military organization that had resources for this. It went much faster after that. The first housing my family received in Chernogolovka was in a four-apartment building. Each apartment had its own entrance and a porch, which was unusual for the Soviet era. Next, they built a small hotel, which was necessary, since many employees still commuted from Moscow.

If Semenov didn't plan to build a town in Chernogolovka, was it Dubovitskii's accomplishment that a town was constructed after all?

Back then, it was extremely important to get a good construction organization to work for you. Once they secured that, it opened up new opportunities for them. For example, the housing situation was challenging at the time, especially for young employees. Semenov and Dubovitskii knew about that. So they decided to build housing for new employees in Chernogolovka, which allowed them to recruit talented employees and expand the scope of the research carried out at the institute. They could also build new lab facilities. At some point, Semenov invited Mstislav Keldysh, a new president of the Academy of Sciences, to come to Chernogolovka. He showed Keldsyh around, demonstrating that there was already a construction organization in place. Keldysh liked that and decided to expand the testing ground and build new research institutes in Chernogolovka. It was around 1961–1963. They established the Institute of Solid State Physics and the Institute of Theoretical Physics. Keldysh was also interested in building a factory for designing scientific equipment.

When did you move to Chernogolovka?

I first came on June 15, 1960.

It's impressive that you remember the exact date!

Well, you see, my son's birthday is on June 16, and he was one day short of being one year old when we moved here.

Did you bring your family with you?

Yes, I brought my family with me, and we received a separate apartment right away.

Did they make you a head of laboratory?

All four of us were confirmed as heads of laboratories in December 1960. You see, they initially planned to expand the work of the four laboratories of the Moscow ICP. They even began recruiting personnel for these laboratories in the late 1950s. But as time went on, it became more complicated to move these labs to Chernogolovka. This is why Semenov decided to create new laboratories at the ICP Branch and to put young people in charge—Manelis, Merzhanov, Dremin, and me. In the summer of 1960, all four families relocated to Chernogolovka. My family arrived first. The other three families came shortly after that. I know the Scientific Council at the Academy had to vote for our appointment, then the Presidium of the Academy of Sciences had to confirm all four appointments.

What did you think about this promotion?

Not much.

How old were you at the time?

I was thirty-one years old.

And what was an average age of a lab director?

It was usually mid-to-late forties. Typically, one could have their own lab, but the time they turned fifty years old.

Did you feel that you were up for a challenge?

One cannot establish a lab in a day, of course. It is a gradual process. At first I had five or six people working in my lab, and Alfred Apin, my advisor at the Moscow ICP, was helping me shape the direction of our scientific research. For a while, he was considering moving to Chernogolovka, so he was helping me get the scientific research started. At its peak, my laboratory had forty people.

What did Chernogolovka look like at the time?

It was basically a village. It was an old village, which served as hunting grounds for Tsar Alexei Mikhailovich in the seventeenth century. When I first visited the site in 1956, there was nothing here but piles of construction materials.

Had your laboratory building already been finished by the time you moved to Chernogolovka in 1960?

It was partially finished. They first built lab facilities for Manelis and Merzhanov, then they built my lab. We only moved to Chernogolovka in 1960, though, because we were waiting for them to finish our two-storied residential cottages.

Were two-storied apartments unusual for the Soviet Union?

The idea to build them belonged to Semenov. His friend Petr Kapitsa had worked abroad in England for a long time. He was well respected in England. Then the Soviet government offered him a position in the Soviet Union and built the Institute of Physical Problems for him in Moscow. But Kapitsa had gotten used to a certain standard of living, so when the Soviet authorities built the institute for him, they also built English-style housing in the institute's backyard. Semenov saw these houses, and since he liked them, he decided to replicate them at the testing ground. This housing was distributed among the institute's leading scientists. At first, they built two cottages, then eight more. Of course, the design was Soviet, not British, and the apartments were not as comfortable. But when a person had no housing at all and received an apartment like that, it was wonderful. Dubovitskii visited us a couple of times. We lived there for about five years until in 1965 Dubovitskii decided to build different, even better housing for us.

You mentioned Petr Kapitsa and his institute. As far as I understand, in 1934, Kapista was forced to stay in the Soviet Union against his will. This was not a voluntary decision, and he was frustrated about it. Recently, I read Kapitsa's correspondence with Semenov from the mid-1930s. In his letters, Kapitsa wrote that a scientist should be allowed to work in any country, as long as his research advances science and scientific development in general. Semenov objected to this, arguing that Kapitsa needed to advance Soviet science and contribute to building Communism in the USSR. What do you think about this exchange?

Kapitsa didn't take into account that the world was divided into two camps: capitalist and socialist. He didn't take into account that there was an antagonism, a struggle between these two camps, and that this struggle could turn into a military rivalry, even into a military conflict at any moment. This is why it was important to advance science in the Soviet Union. Kapitsa didn't take any of this into account.

Do you agree with Semenov then?

Yes, I think Semenov was right.

Do you think it is fine to force a scientist to live and work in a country against his or her will?

This is a complicated question. You see, the Soviet government built an institute for Kapitsa here. He started working, quite successfully, and eventually established his own scientific school. When he worked in England, his research was merely a part of the Rutherford school. As a result, Soviet science began to advance. So from the standpoint of developing science in the Soviet Union, this was a positive development.

But when Kapitsa worked with Rutherford he had his own lab and cutting-edge equipment.

Rutherford thought highly of Kapitsa and later sent his entire lab, with all the equipment, to the USSR. It's true that sometimes scientific research can come to a halt when a scientist leaves. But Rutherford had plenty of scientists working within his school. He could have continued Kapitsa's research even without Kapitsa. Kapitsa was an accomplished scientist, and he made a big difference for the development of Soviet science. If we look at the present day, you probably know that after 1991 a lot of our prominent scientists ended up working abroad. I recently saw an interview with Alexei Abrikosov, for example. He lives and works in the United States now, and feels quite at home there. But the entire direction of scientific research that Abrikosov directed came to a standstill in Russia after he left. It's true that Soviet science lagged behind Western science in some areas, but overall it developed dynamically. Our biggest accomplishments were the nuclear bomb and missile technology.

By the way, I want to brag a little: I am a laureate of the USSR State Prize. The prize was awarded to twelve people, including myself, for "developing the scientific theory of solid rocket propellant." You see, they originally built the testing ground in Chernogolovka to test new powerful explosives. We worked on this for about five years. Then, in the mid-1960s, we received another state assignment: to conduct research on solid rocket propellant. Semenov was eager to get his institute involved, and so we started working on the theoretical foundations for creating solid rocket propellant. It was a challenging task, which required collaboration among numerous research institutes and scientific collectives. We didn't build the rocket, of course, but we conducted theoretical research for it.

Lev Nikolaevich, which political system, in your opinion, provided better opportunities for scientific development?

Scientific development does not depend on a political system. It depends on specific individuals, who recognize the importance of scientific development. It also depends on funding, of course. America is a wealthy country, and they invested much more in science than we did. They issued laws that motivated industry to invest in science. In the Soviet Union, all decisions were made at the very top. Recently, I read an interview with Zhores Alferov in *Pravda*.[15] He said that during his Nobel Prize ceremony in 2000, he participated in a roundtable, together with James Heckman, who was a Professor at the University of Chicago and received a Nobel Prize in Economics that year. And Heckman told him that in the second half of the twentieth-century scientific-technological progress was largely driven by the competition between the Soviet Union and the United States. "Too bad that this completion is over," Heckman added. And he was right: there was a competition between the USSR and the United States in the scientific sphere.

Do you think this competition had a positive impact on the development of science in general?

Yes, it had a positive impact.

Apart from competition in the scientific sphere, did you feel the effects of the Cold War in your daily life?

I personally didn't feel that the Cold War was underway. I remember our newspapers writing that the US government prohibited US companies from selling certain equipment to the USSR, including computers.

Was there hostility toward American scientists, or Americans in general, among the Soviet scientific intelligentsia?

Neither I nor my colleagues felt any hostility toward Americans.

Did you have any collaborative scientific projects with them?

Well, you should remember that for a while our testing ground was a control-access enterprise, and we were not allowed to have any contacts with foreigners. Only in 1975, did I go abroad to Poland for the first time. I didn't travel anywhere else. It is not because they didn't allow me to travel. I simply didn't feel the need to travel. Even during détente, my research remained classified, and I could not report my results abroad. I still read the literature in my field, of course.

Could you read scholarly literature in English?

Yes, but not well. My German was much better.

Did you start traveling abroad more in the 1990s?

Well, Manelis, Merzhanov, and Dremin wanted to travel, but I didn't. My work was also the most classified.

Was such secrecy burdensome for you?

Overall, I was content with the fact that my Soviet colleagues knew about my research, and I didn't worry too much about American scientists. I also had some nonclassified research.

Did you travel abroad as a tourist?

I went to Poland as a scientific tourist, and I paid for it out pocket.

Where in Poland did you go?

I went to Czestochowa. It is a religious center in Poland, similar to Sergiev Posad in Russia. There is a famous monastery there, which is home to the Black Madonna of Czestochowa. There was a five-day conference there, but we stayed longer, for about ten days. We went to Katowice and Krakow. We visited a mine, where we were treated to a typical miner's meal: pea soup with smoked sausage, coarse bread, and a bottle of beer.

Did you try to travel anywhere else, besides Poland?

Yes, I wanted to go to Czechoslovakia, but I was not allowed to go.

And what was the usual procedure for scientists who wanted to travel abroad?

All trips were organized by the Presidium of the Academy of Sciences. They put together a delegation and allocated funding for the trip. Then the party committee held orientations for us, to make sure that we were "politically literate": for example, we needed to know about the political structure of a country we were going to visit. Sometimes a person didn't know these types of things and then he or she was not allowed to travel.

Why did a scientist need to know such things in the first place?

Scientists had to be politically literate. These orientations were also good for our general education. The goal was to make sure Soviet scientists would be able to compare and contrast certain issues, say, regarding Poland and the Soviet Union, in case someone asked them about it.

Do you know if there were any informants traveling as part of your group?

I knew that everyone who traveled with me to Poland were scientists. I am not aware of any of them also being informants.

Lev Nikolaevich, did you participate in any community work at the institute after you moved to Chernogolovka?

I was a head of the institute's party organization for a couple of years. I planned various events to educate Communists at the institute. I invited lecturers from Moscow and organized ideological seminars. For example, we had a seminar on the foreign policy of the Soviet Union or a seminar on political economy. We invited specialists from Moscow, distributed readings ahead of time, and then met about twice a month to discuss them.

Was participation in these seminars mandatory for all employees?

It was mandatory for party members. Sometimes nonparty members joined us as well. Once I remember an employee decided to travel abroad. She was invited to attend a meeting of the institute's party committee, where they asked her if she had been attending any seminars. She said no, and they didn't recommend her for foreign travel.

And did you follow politics?

Yes, from an early age. It was hard to subscribe to a newspaper when I was a child, but there was a newspaper stand near my home. I would go there, stand in line, and buy newspapers. Sometimes, I would leave school during a long break in order to buy *Pionerskaia Pravda*.

How reliable do you think these newspapers were?

I didn't think about these sorts of things. How could I know if the information in newspapers was reliable?

Were you aware of Soviet propaganda?

I started thinking about such things only when I was already a university student. But I didn't have time to worry about politics much. It was challenging to be a student at the Physical-Technical Department, and I simply had no time for this.

After you graduated and started working, did you look for any alternative sources of information, besides newspapers? Did you listen to the Voice of America, for example?

My family was poor and we didn't have money to buy a radio set. I didn't get a radio set until I moved to Chernogolovka.

How old were you?

I was in my thirties.

Did you listen to any foreign broadcasting, such as the Voice of America, after you had bought a radio set?

No, why would I do that? I was not interested in that. I knew we had a lot of problems. Even if I listened to foreign broadcasting and learned about more troubles in the Soviet Union, what could I have done with this information? There was no alternative anyways. What good could come out of it?

Do you remember how your colleagues at the ICP Branch responded to the Soviet invasion of Czechoslovakia in 1968?

We didn't discuss it here.

Did the scientific intelligentsia support the invasion?

People responded differently. Some were skeptical, others thought it was the right thing to do. There were also people like me who believed that there were forces interested in breaking up the socialist camp. They were behind the commotion in Czechoslovakia, because it all came to an end after our troops arrived.

Did you change your mind in the 1990s, when more information became available about Alexander Dubček's reforms?[16]

I still believe that if we had not interfered in 1968, it would have been the end of socialism in Czechoslovakia. Dubček's reforms would have led to the end of socialism, just like we saw in the Soviet Union several decades later. Besides, if there is unrest anywhere, the West is usually behind it. Things like this do not start spontaneously.

Do you think the Soviet Union had the right to tell the people of Czechoslovakia how to live their lives?

We didn't have to tell them how to live their lives. But I think that socialism is better suited for ordinary people than capitalism. I have an example for you. When I was a freshman at the Institute of Fine Chemical Technology, we still had ration cards. There was a canteen not far from the institute, where I could get 100 g of bread, a bowl of soup, and a meatball for one ration card. My friend and I used to go there all the time. Once a young man sat at our table. He had an accent and said he was from Czechoslovakia. We started talking and this young man told us how much he appreciated that in the Soviet Union an ordinary worker could easily approach a main engineer to discuss his work. Moreover, an ordinary worker could marry the daughter of a main engineer. He said that this was unthinkable in his country. We didn't discuss politics at all, we just talked about everyday life. But all my life I believed that a socialist system created more favorable conditions for the working class. Another example I can give you: in the Soviet Union, a janitor and, say, a factory's director could live in the same residential building. Nowadays, when we live under capitalism, you can see that there is housing for the elites, and there is housing for everyone else. There is increasing social stratification.

Some of your colleagues told me the opposite. They said that under socialism people lacked incentives to work.

That is a feature of the Stalinist regime, not socialism. Stalin considered himself to be the most intelligent and literate man in the county, and he wanted to dictate everything to everyone. But such an approach simply doesn't work in the scientific profession, where you need to have your own ideas and initiatives. Stalin misunderstood human psychology.

What about the post-Stalinist period? If socialism is better suited for ordinary people, as you say, why was the economic situation in the Soviet Union so dire? After all, there were constant shortages of food and consumer goods in the 1960s, 1970s, and 1980s.

Socialism had nothing to do with that. That was a consequence of policies introduced from above. Our agriculture, for example, was in bad shape. This was a result of Stalin's policies, when the state and collective farms robbed peasants of all their produce. After the war, they continued on the same course. They didn't invest enough in agriculture. I remember when Khrushchev came to power, he suggested a plan to develop the chemical industry. This plan was confirmed at the party congress. Then all of a sudden, Khrushchev remembered that they also needed to invest in fertilizers for agriculture, and ordered a revision of the plan in favor of producing more fertilizers. But he should have included this in the plan from the beginning.

So did everything depend on the leadership of the country then?

Yes, everything depended on the leadership.

In 1985 Gorbachev came to power and launched a series of reforms. What did you think of perestroika?

Gorbachev was all about words and no action. He said many smart words, but he didn't fulfill any of his promises. Then there was a fight between Yeltsin and Gorbachev. Gorbachev was initially a head of the Stavropol city party committee until Andropov transferred him to Moscow. In 1985 Gorbachev became the General Secretary of the Central Committee of the CPSU. Yeltsin was jealous. Later Yeltsin was appointed the first secretary of the Moscow city party committee. He was responsible for creating food shortages in Moscow.

By the way, what did the food supplies look like in Chernogolovka?

At first, in the early 1960s, there were no grocery stores in Chernogolovka, and we had to bring all food from Moscow. Later, we organized a food parcel system: several employees were in charge of collecting money and going to special distribution centers in and around Moscow. Later Dubovitskii had a small grocery store built in one of the institute's laboratory facilities.

What could you buy there?

It depended on what they brought to the store. I remember there was black and red caviar in barrels at one point. There was also bread, butter, sugar, and meat.

Did this store last long?

Until they built a large grocery store—"Gastronom."

Lev Nikolaevich, do you know if there were any dissidents in Chernogolovka?

What do you mean by "dissidents"? If a dissident is someone who openly challenges Soviet power, then yes, there was one. They put him on trial.

Are you talking about Kronid Liubarskii?

Yes, Liubarskii. He openly criticized the Soviet government. Liubarskii taught astronomy at a local school. By the way, he is the main reason why my daughter decided to be an astronomer and went on to study astronomy in college.

Was he a good teacher?

I don't know. I didn't know him personally. I first saw him in court, when he was already on trial. Dubovitskii asked me to go as an observer, to represent the institute's party organization, so to speak.

What did you think of the trial?

I didn't like it, of course. It was clear that. . . . He was charged with spreading anti-Soviet literature. I think there was a journal published abroad—I don't remember its name now—and somehow Liubarskii received this journal and distributed it.[17] You see, one could own foreign journals and read them at home, this was not a crime. But when you started giving this literature to other people to read, you became a distributor of anti-Soviet literature. And that was the main charge against Liubarskii.

Did you or anyone among your friends or acquaintances borrow literature from Liubarskii?

No, no one I knew borrowed any literature from him. At least they didn't tell me about it. I guess he had his own circle of people who did borrow this literature, but those people didn't distribute it.

Was it just Liubarskii who was put on trial then?

Yes.

Was he found guilty?

Yes, he was sentenced to several years in prison. Later a KGB agent came to our institute in Chernogolovka. He told us that when Liubarskii's sentence ended, and they set him free, they said: "We hope you won't end up here again." And he responded: "It depends on how well you do your job."

Was he planning to continue his dissident activity?

Yes, he was planning to continue.

Do you think it was a right thing to do?

It was his personal choice.

Yes, but do you think distributing this literature was beneficial or harmful to the Soviet state?

How could it be beneficial?

Do you think Liubarskii's activity was harmful to Soviet society?

A word can be used as a weapon, too. If this word targets a government, it is a political weapon. And if this word becomes accepted by the majority of people in a given society, it can bring social change. Then the question is whether this change would be for the good of the country or not. This is what was at stake: freedom of speech, free elections, political alternatives, and so on. But you see, if authorities had granted us freedom of speech, this would have become a weapon for opponents of Soviet power. What would be the potential consequences of this? It would have led to the situation we have today.

In the 1990s, I sometimes talked to my younger colleagues, and they all criticized the government and said that this or that had to be done differently. They wanted change, but they didn't understand human psychology and didn't know how to bring about positive social changes. It is an extremely complicated question. What could be done in the Soviet context? Let's imagine that the Soviet government agreed to form an alternative political party and allowed it to compete with the Communist Party. Let's say this would be a peasant party. What could that party have done? I suppose they could have redistributed land that belonged to collective farms. But what would the end result be? It would have led to the abolition of collective farms and redistribution of land among peasants. They did something like this during perestroika. At the end, peasants slaughtered all their cows because they could not afford feeding them. Collective farms, by contrast, had tractors and other equipment.

You see, the Communist Party said that it served the people. Of course, it didn't always deliver. But even if there had been an alternative political party in the Soviet Union, what could it have done differently? I simply cannot image that. By the way, at some point, Khrushchev tried to divide regional party committees into agricultural and industrial.[18] He also carried out the Virgin Lands campaign and promoted the expansion of corn cultivation. Both were ill-conceived ideas. Some people praise Khrushchev for denouncing "the cult of personality." They say that this was Khrushchev's most important accomplishment. But did his "secret speech" make anyone's life easier?

Well, Khrushchev put an end to the mass repressions. People were no longer arrested randomly, in the middle of the night, and sent to concentration camps. Would it have been better if Stalin had been succeeded by a leader who was like Stalin?

This could not have happened. Although Gorbachev aspired to be like Stalin.

In what sense?

He also wanted to tell people what to do and he was intolerant of criticism coming from the Central Committee. People who worked with Stalin say that even he was open to serious discussions. His only condition was that people who argued with him were experts in their areas. Stalin didn't admit his mistakes, of course, although he made mistakes, especially during the war.

Lev Nikolaevich, changing the subject, do you think there was more intellectual freedom in Chernogolovka than in the rest of the USSR?

Never in my life did I feel that I was not free intellectually. So it is hard for me to answer this question. When it came to science, I was always able to say whatever I needed to say. Besides, I was Dubovitskii's deputy director.

When was that?

I was his deputy director from the late 1960s to the late 1970s. And although Dubovitskii was a rather authoritarian man, I was not afraid to disagree with him. I should tell you, though, that I really didn't like it when people told jokes and anecdotes about Lenin. Some people believe in God, others believe in Communism. I believed in Communism. And since this was my faith, Lenin was a saint for me.

Are you saying that Communism is a kind of religion that requires faith and that it is not just another way to organize society?

You see, we never lived under Communism, and we don't really know what it is. Today it is simply a faith.

But what is Communism exactly?

I don't know.

You just said that you believe in it.

I believe that one can organize a society in such a way that people would live according to certain moral and legal laws, where people would help each other instead of living in a dog-eat-dog world.

What are the moral laws of a communist society?

Do not harm another man.

And how can one reconcile this with the mass repressions in the Soviet Union?

One cannot.

And if Lenin was a saint, why did he persecute his political opponents?

Who did he persecute? You need to understand that there was a class struggle underway, and everyone was fighting for their own interests. These interests were impossible to reconcile. This is what led to the Civil War. Did he shoot his own people? Not really.

He launched repressions against members of other political parties, including the Left Socialist Revolutionaries (SRs), immediately after the October 1917 Revolution.

The left SRs organized a mutiny against the Bolsheviks, because they disagreed with the policy of War Communism. This mutiny had to be suppressed. You should keep in mind that Lenin was an emotional person. He said many things, but he didn't mean all of them.

What about the establishment of the first labor camps? What was their purpose?

When were they established?

The Solovki special camp was set up in the early 1920s.

Lenin was out of office by 1922. It was Trotsky who took a much more assertive position.

NOTES

1. *Rabfak*, or worker's school, provided workers and peasants with general education, preparing them for university entrance. The idea was to make higher education more accessible.
2. Vashin is referring to the Winter War of 1939–1940, during which the Soviet Union invaded Finland.
3. The GOELRO (State Commission for the Electrification of Russia) Plan was launched by Vladimir Lenin in 1920 and was the first plan for bringing electricity to all of Russia. The Soviet government portrayed it as a crucial step in building Communism in the new Soviet state.
4. Arzamas-16, present-day Sarov, was the epicenter of the Soviet nuclear program.
5. The Central Aerohydrodynamic Institute (TsAGI) is a research institute in Zhukovsky, a city located twenty-five miles southeast of Moscow. Founded in 1918, TsAGI specialized in aircraft design and construction.
6. VKP(b) stands for the All-Union Communist Party (Bolsheviks). In 1952, the name of the party was changed to the Communist Party of the Soviet Union (KPSS, or the CPSU).
7. The collectivization of agriculture was part of the first Five-Year Plan (1928–1932). The campaign proceeded in several stages, including "dekulakization" and deportations. The term "kulak," originally used to describe the better-off peasants, was in effect applied to anyone who resisted joining the collective farms.
8. Kolkhoz was a form of collective farm in the Soviet Union.
9. *The Wizard of the Emerald City* is a children's novel by Aleksandr Volkov, published in the USSR in 1939. It is a re-narration of L. Frank Baum's *The Wonderful Wizard of Oz.*
10. The anti-cosmopolitan campaign targeted the Soviet Jewish intelligentsia in the late Stalin era.
11. The monetary reform of 1961 introduced a new ruble currency at the rate of 1 new ruble to 10 old rubles. Effective January 1, 1961, all prices, wages, and other monetary incomes were also reduced to one-tenth of their former level.
12. The Soviet partisans were an underground resistance movement in the USSR during World War II.
13. The economic policy introduced by the Bolsheviks during the Russian Civil War (1918–1921). It included nationalization of trade and industry, suspension of a

market mechanism, and partial suspension of money. The policy led to the devastation of the national economy. It was replaced by the New Economic Policy (NEP) in 1921.

14. An equivalent of a doctorate degree, or PhD. It was the first of the two doctoral scientific degrees in the Soviet Union.

15. Zhores Alferov (1930–2019) was a prominent Soviet and Russian physicist and a winner of the 2000 Nobel Prize in Physics. Later in life, he also became involved in politics, serving in the State Duma as a member of the Communist Party of the Russian Federation.

16. Alexander Dubček was the First Secretary of the Communist Party of Czechoslovakia. In April 1968, he launched a series of reforms, which called for more openness and partial economic decentralization in Czechoslovakia. The reforms led to the Soviet invasion of Czechoslovakia in August 1968.

17. Kronid Liubarskii helped produce and distribute *The Chronicle of Current Events*, an underground journal produced in the Soviet Union. It published information about political repressions in the USSR.

18. In 1962, Khrushchev decided to split party committees at the local level into agricultural and industrial sections. He hoped that it would make the administration at the local level more efficient. However, the reform was a failure and contributed to Khrushchev's removal from power in 1964.

Chapter 5

"I Began to Critically Evaluate Soviet Life after I Met Kronid"

Rustem Bronislavovich Liubovskii

The following interview with Rustem Bronislavovich Liubovskii (b. 1938) is a testimony to the gradual disillusionment with Communism among some members of the Soviet scientific intelligentsia. A soft-spoken and inquisitive man, Liubovskii was one of the few residents of Chernogolovka who befriended the dissident scientist, Kronid Arkadievich Liubarskii. Liubarskii moved to Chernogolovka in 1967, following an invitation to teach physics and astronomy at a local school. By then he was an active member of the dissident circle in Moscow. Rustem Bronislavovich and Kronid Arkadievich translated Chemical Evolution, *a book by Nobel Prize laureate Melvin Calvin. They also spent many Saturday evenings at Liubarskii's apartment, discussing politics, current affairs, physics, and samizdat texts, which Rustem Bronislavovich had borrowed from his dissident friend. The reading and distribution of samizdat was proclaimed illegal in the USSR and was persecuted by Soviet authorities. Liubovskii borrowed it anyways, since reading samizdat gave him access to information that was not available in official Soviet sources. Liubovskii's friendship with the dissident broadened his horizons and eventually made him question the very foundations of the Soviet system.*

Unfortunately, the friendship didn't last long. In January 1972, Kronid Liubarskii was arrested after a KGB search of his apartment in Chernogolovka. He was accused of spreading samizdat literature, put on trial, and sentenced to five years in a labor camp. Rustem Liubovskii was summoned to the KGB for an interrogation several months later. For a while, he feared that he would lose his job at the Branch of the Institute of Chemical Physics and would not be able to support his family. Even though Liubovskii did keep his job, his first-hand encounter with the repressive side of the Soviet system left him exceedingly uneasy. He spent the next two decades searching for answers and truthful information outside of the official Soviet narrative.

I first met with Rustem Bronislavovich in late December 2009. At the time, I was collecting information about the 1972 trial of Kronid Liubarskii, the so-called Liubarskii Affair, and was conducting interviews with Chernogolovka scientists. While most of my interviewees sympathized with Soviet dissidents, they were reluctant to discuss what had happened to Kronid Liubarskii. Over and over again, they kept telling me that I had to talk to Rustem Liubovskii instead, since he would know much more about the Liubarskii Affair. Rustem Bronislavovich agreed to meet with me right away and gave me several interviews over the course of the next several months. In his interviews, Liubovskii talked about the pervasive fear in Soviet society, which passed on from one generation to another. He insisted it was crucial to conquer that fear for the healthy development of post-Soviet Russia. It appears that Liubovskii has spent most of his adult life, especially after Kronid's arrest, striving to achieve that goal.

The story of Rustem Liubovskii is not an account of another Soviet dissident, however. In many ways, his story is typical of many members of the postwar generation of Soviet scientists. Born in 1938, Rustem Bronislavovich was a child during the Great Patriotic War, which he remembers as "an extremely difficult and hungry time." His parents moved around a lot: from Arkhangelsk to Tbilisi to Gagra, then Sukhumi. After the war ended, Liubovskii's father, who had served in the First Polish Army, brought his wife and two children to Poland, which was Liubovskii's first trip abroad. In 1947 the family returned to the Soviet Union. They chose to settle in Lviv, a city in western Ukraine, which had been recently annexed by the USSR. Rustem Bronislavovich attended middle school and high school there. In 1957 he left Lviv for Moscow, where he successfully enrolled in the prestigious Moscow Physical-Technical Institute (popularly known as Fiztech*), the Soviet equivalent of the Massachusetts Institute of Technology. After graduating from Fiztech in 1963, Liubovskii accepted a job at the Branch of the Institute of Chemical Physics in Chernogolovka, where he has worked ever since.*

Before he met Kronid Arkadievich, Liubovskii was not really interested in the politics or history of his country. During the interview, he recollected that his paternal grandparents were small business owners, who had lost their house and their mill after the Revolution of 1917. In the 1930s, Liubovskii's father was repressed as well. Despite this, Rustem Bronislavovich grew up to embrace the Soviet way of life and to internalize the official Communist ideology. "We were all raised to think the same way," he explained in one the interviews. Not even Khrushchev's "secret speech" in February 1956, in which the first secretary denounced Stalin's crimes, cast any doubts on Liubovskii's worldview. He remained largely apolitical during his student years at Fiztech, where he spent most of his time studying physics. He maintained his apolitical stance well into the late 1960s, preoccupied with science, starting a family, and organizing various cultural events in Chernogolovka.

Liubovskii's friendship with the dissident fundamentally changed his views on the Soviet state and society. It made him into a vehement anti-Stalinist. In the interviews, he called Stalin "a dangerous, dreadful person." He held him responsible for the high price that Soviet people had to pay for industrialization, as well as Soviet victory in the Great Patriotic War. Conversations with Kronid, as well as reading The Chronicle of Current Events, *which published information about political prisoners in the USSR, opened Liubovskii's eyes to the repressive side of the Soviet system. Decades later, he was especially appalled by the regime's lack of respect for the individual. The transformation of Liubovskii's views didn't happen overnight, though. It was a slow and gradual process. He never fully accepted the dissidents' bleak portrayal of Soviet reality, pointing out that there had been tangible improvements in the USSR after Stalin's death. At the same time, Liubovskii's account is remarkably devoid of nostalgia for socialism or the Soviet era.*

Rustem Bronislavovich, when and where were you born?

I was born on September 30, 1938, in Tbilisi, even though in my passport it says I was born on September 30, 1939, in Volgograd. Later, my mother told me that she had changed my date and place of birth after the war, when she applied for a new birth certificate for me. There was compulsory military service in the USSR, which started immediately after high school, and she wanted me to have an additional year for college applications. She put down Volgograd, or former Stalingrad, as my place of birth, because most of the city had been destroyed during the war, which meant that many records had been lost.

My family moved around quite a bit. During the war, we lived in Gagra, a town in the Caucasus. It was an extremely difficult and hungry time. People were starving, because even though the Germans didn't occupy the Caucasus, their troops laid siege to the region. They were trying to get to Grozny and Sukhumi, but failed. Yet they managed to cut off supplies to the North Caucasus, creating starvation in the region. We only survived thanks to fruit that grew in the south: figs and grapes. Our house was also located next to a bigger road, and sometimes people in passing trucks threw sunflower cakes at us children, and we were very grateful. I was five or six years old at the time. I remember one day I climbed up a big fig tree to pick figs, and I fell and hurt myself badly. My father was on the frontlines. My mother was busy raising my little sister and me. So when I got home, I didn't tell anyone about the fall. But six months later I began limping. It turned out that the fall caused the beginning of tuberculosis. This was in 1943. My mom took me first to Sukhumi to get the X-rays done. Then she checked me into a health center in Tbilisi, where I stayed until the end of the war. I had to wear a cast for a while, which led to some additional medical problems.

By the way, my mom was Russian, but my dad was Polish. Yes, my father was Polish, even though I always wrote that I was Russian in official documents. I think I have the right to call myself Russian: I lived here all my life. Anyways, when the war broke out, my father joined the Soviet Army. He ended up serving in the First Polish Army under the command of General Berling. In 1945, his army reached Warsaw, and since I was still sick, my father decided to bring me, my mom, and my sister to Warsaw, where I could get better treatment. I was seven or eight years old, and it was my first time abroad. I remember seeing Warsaw for the first time. Most of the city lay in ruins. The old town had been largely destroyed. My dad took me to the top of some building, and I could see that the only surviving structure around us was a Catholic church. In Poland, I spent a year and a half in a health center in Zakapane, which nowadays is a famous skiing resort. Other children teased me and called me "*moskal*,"[1] because I spoke Russian. When my parents came to visit me three months later, I asked them to speak Polish to me.

Did your parents speak Polish?

My dad knew Polish, of course. My mom and my sister didn't learn it until they lived in Poland. I learned it quickly, too, in three months or so. Then in 1947 Soviet policies toward Europe changed. Actually, they started to change in 1945, with the outbreak of the Cold War, which brought a new division of the world. Since the Soviet government decided to withdraw some of its troops from Poland, my parents had to choose whether to stay in Poland or to go back to Russia. They decided to return to Russia, and in 1947, my family moved to Lviv.[2] I am not sure what guided their decision. In Lviv, I went to second grade, where local children teased me again, calling me "*pshek*" this time, which means "a Pole." After I finished school, I applied to the Lviv Polytechnic Institute, but didn't get in, fortunately. I say fortunately, because the following year I went to Moscow and passed the entrance exams at the Moscow Physical-Technical Institute, or Fiztech for short. I graduated from Fiztech in 1963 and moved to Chernogolovka. This is a brief overview of my professional life.

What did your parents do before the war and did their occupation change after they moved to Lviv?

My father was a baker. He was a talented baker, too, one of the best ones in Lviv. I still have photographs of some of the cakes he made. They looked like artwork. When Rimma and I were getting married, my father baked us a large cake, decorated with pigeons. Why did he become a baker, you might wonder? In 1935 my father was exiled to Arkhangelsk. Before the exile, he was a student at an institute in Kiev, but he said or did something, which got him expelled from the institute. I don't know exactly what happened. My father was never a political person, but those were dreadful times: one imprudent word could get you in trouble. I am getting a bit off topic now, but

Stalin's birthday was several days ago. There were a lot of programs on TV about it. Oftentimes, I hear people say that Stalin was a strong leader and that Russia needs one today. And I always think to myself: "No, it doesn't!" Stalin was a dangerous, dreadful person. "But he saved Russia during the war," I hear people say. No, he didn't save Russia—the Russians did! Stalin was in charge, of course, but what price did he have us pay for our victory? It was too high and brutal. It involved too much bloodshed and physical destruction. Stalin didn't appreciate the value of human life. He saw individual people as cogs in the state machine. It was terrifying.

Some people today also praise Stalin for carrying out industrialization. They say he modernized the Soviet Union in a short span of time.

True, but this could have been done differently. If industrialization had been carried out more sensibly, it would have had much better results. Besides, I think that if it had not been for Stalin, we could have avoided World War II altogether. Stalin provoked that war, when he threatened to start a world Communist revolution and to spread Communism around the globe. And what about the expulsion and the physical annihilation of the Russian intelligentsia in the 1920s? It was a thin layer of society, yet these were some of the most educated and knowledgeable people. Every society depends on this thin layer for its development. Of course, there were a large number of people at the bottom, who wanted to rise up on the social ladder.

Sometimes you hear today that this is what a capitalist system looks like: the strongest rise to the top, while the rest become their workforce. I am not sure how accurate this description is. I do know, though, that it is important to have normal human relationships both at work and in everyday life. People should have their rights and responsibilities. If someone made a lot of money, they should share with the rest. It is extremely difficult to get there. Capitalism is a good system overall. I read somewhere that capitalism allows individuals to do what they can do best, without anyone's interference. This means that people can provide for themselves and their families. On paper, socialism might look even better. "From each according to his ability, to each according to his needs." It sounds good in theory, yet what does it teach us? It teaches us to be lazy and to consume without being productive. This is just the way human beings are. Socialism destroys motivation. When people came to work in the Soviet Union, they used to ask for detailed instructions, instead of taking initiative. This was one of the tragedies of Soviet socialism. Under capitalism, you must find work and prove yourself worthy, instead of simply waiting for someone to tell you what to do.

Another problem in the Soviet Union was the low levels of self-awareness. As a result, ideology replaced critical thinking. This allowed Stalin's dictatorship to flourish. Even in the midst of mass repressions, people didn't know much about what was going on. Both of my parents, for example, were afraid to talk about the repressions, even though my father himself had been repressed. Moreover, my parents didn't tell me much about their families' background. For a while, I didn't know anything about my father's side of the family. I can only suspect that they were small business owners. As for my mom, she was born and raised in Arkhangelsk. This is where she eventually met my dad. Her father was a Russian Cossack, who perished around 1917. Her mother was German. I know they spoke German in their household before the Revolution, but it became dangerous afterward, so they started speaking Russian instead. Many people had to change who they were to fit into the new system.

This is all I know about my family's background. I wish I knew more because nowadays my children keep asking me about their ancestors, especially prior to the 1917 Revolution. But my parents were afraid to tell me who their parents were and what they did. I didn't ask many questions growing up either. Currently, I am trying to gather some information from a few surviving relatives. I managed to find out, for example, that my father's parents had a mill before the Revolution. They lived in Zhytomyr, in Western Ukraine. My father came from a large Polish family: they had ten children. My grandmother didn't work and took care of the children. My grandfather worked at the mill. In the Soviet era, we had to fill out various forms, which included questions like: "Were you or any of your relatives prisoners of war or did they live in the occupied territories? What did you do before the Revolution?" People were terrified of these questions, since they could get them into trouble. My parents chose not to even tell me about my ancestors to spare me.

Do you know what happened to your father's family after the Revolution?

They lost everything: their house and their mill. My dad was born in 1907. He studied in Kiev, but as I already mentioned to you he didn't finish his education. He was Catholic and he helped with the service at a local Catholic church. Maybe this is why he was repressed. I don't really know what happened there, but he was expelled from his institute and exiled to Arkhangelsk. This is where he met my mom, who was several years his junior. As far as my mom's side of the family goes, she was one of six children. Her family lived in Arkhangelsk. I still don't know what my maternal grandfather did. I actually never met either of my grandfathers: they both died early. I did know my grandmothers, though. They lived until their mid-nineties. You see the relationship between our generation and our parents' generation was broken. Our parents worked extremely hard to support their families, but had little time left to spend with their children. They were also afraid to talk to us about certain things. When I applied to Fiztech, for example, they checked my background. If they had found out that I had a previously repressed parent, it could have caused serious problems.

How long did your family live in Arkhangelsk?

Not that long. My sister was born in 1935. I was born three years later. At that point, we already lived in Tbilisi, where my father got a job. Then we moved to Sukhumi. In 1939 my father was arrested. Someone had denounced him. Thank goodness, he didn't stay in prison that long. My mom was an energetic woman, and she started writing letters to the prison's administration, demanding to know why her husband was in jail. Eventually, they let him go. I was only one and a half years old. After that, we moved to Gagra, where we lived during the war. In Gagra my dad worked at a cake factory until he was enlisted in the army. We stayed in Gagra, while he was on the frontlines. My mom, my sister, and I lived in a small room. We barely had anything to eat, even though my mom worked three jobs.

Did your mother have a formal education?

No, she finished some courses, but didn't have any formal education. She married early, in 1934, when she was still a young girl. I remember she once told me that she wanted to be an actress and that she even had applied to some institute in Leningrad. My mom was beautiful and she played guitar well. Her application got rejected, allegedly because she made several grammatical errors. So she finished some courses instead and became an accountant. During the war, my mom worked three jobs, and we barely saw her at all. This was not unusual, though, since the war was underway. After the Germans laid siege to Gagra, they caused a famine inside the city. I already told you how I climbed up the tree one day, fell, and hurt my leg. After the war, my family went to Poland, trying to save me and treat my tuberculosis. Once I got better, we moved to Lviv.

Did you parents discuss politics or what was going on in the country at home?

You know, we barely talked to our parents. In Lviv, my dad always left for work early in the morning and came home late at night. My mom also worked long hours. For a while, we all shared a small room in a communal apartment in Lviv. Later, around 1950, we received two rooms in another communal apartment—all thanks to my mom's efforts.

What else do you remember about the war?

I remember hunger. One time my friends and I went into the forest and found some nuts there. We ate them, but they turned out to be poisonous. No one died, but we got really sick. Another time I remember my grandmother made some fruit jam. My mom was at work, so I called my friends, and we ate almost all it. We felt horrible afterward. I also remember how my dad came back to visit us for a couple of days during the war. He had been wounded, and they allowed him to go see his family. It was around 1944. He was wearing a military uniform and looked so handsome. Then we left for Poland, where my family lived from 1945 to 1947, but I already told you about that.

Did your parents consider staying in Poland?

Yes, they thought about staying in 1947, and my dad was inclined to stay. But my wise mom said that we were Russians, and we should return to Russia.

Do you realize that you were one of the few Soviet children who traveled abroad in the 1940s?

Yes, this is true. I attempted to go abroad again later, when I was already working in Chernogolovka. I remember in the 1970s, I applied to travel to Czechoslovakia to go skiing there, but I was denied permission to leave the country. We had a skiing club in Chernogolovka, and I applied to go to Czechoslovakia as part of a tourist group, but they didn't allow me to go. They simply crossed my name off the list. Maybe it was because I had said something critical about the Soviet invasion of Czechoslovakia.

What year was that?

It was 1975–1976.[3] Besides, I had already been implicated in the Liubarskii Affair. So I could forget about traveling abroad for a while.

Why do you think your mother chose to return to the Soviet Union? Were the material conditions in Poland similar to those in the USSR?

They were better in Poland. Moreover, the military men earned good salaries. But my mom wanted her children to get a good education, which is why she decided to return to the Soviet Union. After my family moved to Lviv, my father left the military and got a job as a baker. He always had a couple of sweets for us in his pocket. In a sense, we had a sweet life (laughs). In Lviv I went to second grade, and I finished middle school and high school there.

When you were in high school, did you already know that you would become a physicist?

No, I didn't. I didn't have much direction when I was in school. I only remember that my mom always encouraged me to study hard and be a good student. It was important for her that I got a good education. My sister, who was several years older than me, finished school and went on to study at the Lviv Polytechnic Institute. She became an engineer. When I finished school, I applied to the Lviv Polytechnic Institute, too, since my sister was a student there already, but I didn't get in. It was around the same time that I learned about Fiztech in Moscow. This is how it happened. My former classmate, Robert Minosian, finished school and went to Moscow to apply to some university there. He was rejected and returned to Lviv. When he came back, he told me that he had tried to get into the Moscow Physical-Technical Institute, or Fiztech. I immediately decided that I wanted to apply there as well. Since I had a year in-between, I got a job as a lab assistant at my former high school. I worked there during the day and spent all my free time studying for the entrance exams at Fiztech. I even managed to find some Fiztech textbooks and studied them thoroughly, solving various problems, which I found quite enjoyable. The following summer I went to Moscow and successfully passed the entrance exams at Fiztech. This is how I became a student at the Radiotechnical Department there. I took courses at Fiztech for the first three years. In my fourth year, I also interned at the Kurchatov Institute,[4] where I conducted research on solid state physics.

When did you become a student at Fiztech?

Let's see here. I finished school in 1956, but I didn't get into the Lviv Polytechnic Institute. So in 1957 I went to Moscow and passed the entrance exams at Fiztech. They provided me with a dorm room, where I could live. Since the coursework was extremely challenging, I was completely overwhelmed at first. For the first few months, I could not even understand what some professors were saying. I wasn't the only one who struggled, of course. Only gradually did I feel more comfortable there. I attended Fiztech from 1957 to 1963, so six years overall.

What was it like to be a student at Fiztech?

It was a closed and highly selective institute. Some people failed to get in because they were Jewish. By the way, when I came to Moscow to take my entrance exams, the Moscow Youth Festival was underway. So I spent a couple of nights wandering around Moscow, although I didn't talk to anyone in particular. Anyways, I spent my six years at Fiztech studying almost nonstop. Science was the only thing on my mind. At the time, I had no interest in politics whatsoever. Sure, we had to take some required courses on political economy in order to pass the exams. But I was completely apolitical at the time. Sometimes my friends and I went to a theater or a conservatory. We also invited artists to perform at Fiztech, such as Bulat Okudzhava, Ilya Ehrenburg, and Mikhail Romm. I remember Ehrenburg gave a talk, in which he was comparing capitalism and communism. We were surprised by his frankness. In general, though, we didn't discuss politics as students. We didn't even have a radio. Science occupied all our time.

What were the attitudes toward science and scientists in the late 1950s and early 1960s?

There was a special attitude toward science in Soviet society. Since scientists helped create the bombs and launch Sputnik into space, the state supported science. In fact, the state invested a tremendous amount of money into scientific development. Once we claimed that communism would defeat capitalism, we needed weapons to support this claim. I remember when I was still a high school student, they kept telling us that Russia could destroy America ten times over, while America could destroy Russia only nine times over. In short, we had more weapons than we needed. But the overall attitude toward science was positive. Yet a lot of the funding went toward military-oriented research and the army.

Did Soviet officials respect scientists? Some of your colleagues told me that scientists enjoyed a special relationship with the Soviet regime.

Well, in the Stalin era, Soviet officials persecuted scientists. Many of them ended up working in *sharashkas* in labor camps.[5] In fact, Sergei Korolev, who launched the first Sputnik and put the first man into space, spent some time in the Gulag. This changed in the 1950s and 1960s, of course. The state began to allow scientists a certain degree of intellectual freedom. Soviet dissidents played a large role in this.

Rustem Bronislavovich, let's talk about dissidents in a little bit. What happened to you after you graduated from Fiztech and how did you end up in Chernogolovka?

Well, during my last several years at Fiztech, I interned at the Kurchatov Institute. I knew that I would not be able to find a job at a research institute in Moscow, since I didn't have a Moscow registration. So the head of my laboratory, Samoilov, told me about a scientific center in Chernogolovka. He said that they were recruiting scientists and offered to introduce me to Igor Shchegolev, who was a head of a research group in Chernogolovka. Samoilov gave me Shchegolev's contact information and I got in touch with him. This is how I ended up working at the Branch of the Institute of Chemical Physics.

Nowadays, when we reflect on our lives back then, those of my friends who are still true Communist believers, argue that things were great in the Soviet Union. It is simply not true. Things were not great. We were not free. Moreover, we had this constant internal fear, which we inherited from our parents. I already told you that our parents were afraid to discuss many things with us. We didn't know anything about labor camps, for example. They were too afraid to speak. Even today, when I talk to some of my old friends, I can see that they are still afraid. This old, slavish mentality is still alive. To this day, many of us are slaves to that fear. But what is there to be afraid of now? Most of us are in our seventies and eighties already. Moreover, we are not even aware that we are afraid: this fear has been ingrained in us from an early age. We inherited it from our parents and grandparents. But you should live your life in a way that would not make you feel ashamed later on. Do not be afraid of the word. The word will not hurt—it will tell the truth. Nowadays, I am not afraid of anything, and thank God. But this is the result of the democratization brought about by Yeltsin and Gaidar.[6] Gorbachev contributed to this as well. Many people hate them, but I respect them, because they brought about these changes. It is not their fault that things didn't work out perfectly. We need to take responsibility for our lives as well.

Rustem Bronislavovich, you said that you were not free. Yet some of your colleagues argue that they didn't feel that their freedom was restricted in any way.

This is because we were isolated from the rest of the world. We simply didn't know what was going on beyond Soviet borders. Sometimes, we could catch the Voice of America on our radio set at home. But most people didn't listen to it anyways. Why would they? They were finally able to buy bread freely. Once in a while, they could even buy shoes and clothes that were "thrown out" (*vybrosili*) to us.[7] This was enough to keep people content. But we could not discuss things openly. As soon as you opened your mouth, you immediately got in trouble. There were informers everywhere. They even tried to recruit me once, but I firmly refused.

Of course, we received free housing and free education from the state. Many older people remind you of these benefits when you talk to them today. But at what cost? It was only possible because all of us survived on minimal salaries—100 rubles a month at best. What could we even buy with that money? Besides, there was nothing to buy in stores. We were not allowed to travel abroad either, so it didn't even make sense to save money for that. People tried to save anyways, for a rainy day, as they say. But could we dream of buying a car at some point down the road? Of course not! We could not even afford buying a TV set. In 1965, my wife and I were finally able to buy a refrigerator. It was made in Finland. In fact, we still have it in our country house (*dacha*). Nowadays, when I think about how we lived, I realize that we could only enjoy free housing and other state benefits because overall we had a very low quality of life. At the same time, the government invested a lot of money into the military.

Yet when we compared our lives to the lives of our parents, we could not help but think that we lived in paradise. Indeed, our material conditions were much better than theirs. The question then is what an individual needs more: freedom and the rule of law, or order, even if it comes at the expense of freedom? What is more important? Europe and the United States live by the rule of law. What we have here today is dreadful. There is no rule of law in Russia. It will take several generations for us to get rid of the fear that was ingrained in us, and to start thinking freely. Most of us have that fear deep inside. Even today, when I need to speak in public, I think in the back of my mind about how it might impact my life.

Going back to your first few years in Chernogolovka, what did the town look like at the time?

When I moved here in 1963, there was only one street in Chernogolovka. It was called First Street (*Pervaia Ulitsa*). One side of it had two-storied apartment buildings, and they were building four-storied housing on the other side. I received a room in a communal apartment, which I shared with two other people. There were three rooms in that apartment, and the three of us occupied one of the rooms. I lived there from September to December 1963. In December, they finished another residential building, and my colleague and I got an apartment there.

Did Chernogolovka have good access to groceries at the time?

No, not really. I remember, though, that they sold some red caviar in a grocery store at our institute, shortly after I came here. Of course, it was only because Dubovitskii made special arrangements for us. The caviar was sold in bulk.

What other products were available at the store?

I only vaguely remember. We didn't need much, you know. We could actually live on potatoes alone. Later, they opened a large grocery store in town. I am not sure if it was better supplied. I do remember that when we had children, we needed to buy milk for them, but it was difficult to find.

Did you participate in landscaping Chernogolovka?

Well, yes I did. We had *subbotnik*.[8] In 1963–1964, I helped with the construction of recreational facilities, including tennis courts and volleyball courts. In the mid-1970s, I also directed building a recreational park in Chernogolovka. Dubovitskii knew that I had organized seasonal work trips, or *shabashkas*, to Norilsk, where we earned a lot of money. Dubovitskii scolded me for this. He thought we should invest all our energy in building Chernogolovka. So he called me to his office one day and said that he wanted me to organize the construction of a park on April 21, which was Lenin's birthday. So I gathered people who had worked in Norilsk with me, about several hundred people. We drew a sketch of the park and Dubovitskii supplied us with shovels. But when we came to *subbotnik*, it snowed, which we didn't expect to happen in late April. What was I to do? I didn't want to cancel it. Since we had a previous experience working in Norilsk, we decide to proceed anyways. I broke people into small groups and we started cutting down trees and digging ditches for electric cables. By noon we found out, however, that our institute's party organization had a meeting in the morning and decided to cancel the *subbotnik* due to inclement weather. They didn't know that we had already started construction. Around noon, several people from the institute's administration showed up and began scolding me for violating their order. But since the work was already underway, Dubovitskii had no choice but to order hot tea and meat pie and have it brought to the construction site.

I also actively participated in organizing cultural events in Chernogolovka. When we first came here, we were young and energetic. Since there was no cultural center in the town yet, we decided to take matters into our own hands. So we would go to Moscow and buy tickets to theatrical performances, then distribute those tickets among Chernogolovka residents. We would then reserve the institute's bus to take us to Moscow. We also organized performances in local villages, where we would dance and read poetry. This is how I met Rimma, my wife. We knew right away that we were a great match.

Later we started inviting Moscow artists to come and give performances in Chernogolovka. We also arranged to bring films from Moscow to screen them here. In the late 1960s, I helped Timirbaev, Babkin, and Zvarykina organize a cinema club in Chernogolovka. It was called "Kaleidoscope." At that point, though, we already had children, and I could not spend as much time on that. The cinema club was part of the House of Scientists. It was extremely popular. I remember how we would gather in someone's kitchen and discuss what film repositories to go to. We were also in charge of distributing tickets, which was usually challenging: there were only 600 members of the House of Scientists, but many more people wanted to come to screenings. "Kaleidoscope" presented an opportunity for us to see something new, something that one could not see even in the most liberal Moscow cinemas.

Rustem Bronislavovich, you mentioned that you had organized seasonal work trips, the so-called shabashkas, *to Norilsk and that all participants could earn a lot of money on these trips. Can you please tell me more about them?*

The idea came to us almost by accident. In 1965 Rimma Nikolaevna Stepanova and I got married, and in 1967–1968 we decided to go mountain hiking and camping by the Black Sea. We had to cover about 50 km per day, carrying heavy backpacks with us. It was hard. When we returned home, I complained about this to someone. And they told me that they heard that some students from Fiztech had gone to Norilsk that summer to do some seasonal work there. We didn't really know where we would go and how we would organize our trip. But rumor had it that one could make 600 rubles in about forty days there. Keep in mind that our salaries were only about 130 rubles. Salaries went up to 170 rubles, once you defended your candidate of sciences degree. But an average salary of a junior research scientist was 130 rubles. So an opportunity to make 600 rubles in forty days was extremely appealing to us. We talked it over at home and decided that we should look into it. The following day I mentioned this to someone in my lab at work. The head of my lab was Igor Shchegolev, who was a student of a famous Soviet scientist, Petr Kapitsa. Shchegolev was a wonderful and highly accomplished man, everyone liked him. So I came to work and said: "Fellows, why don't we go to Norilsk? We could make good money and see some new places." One of my colleagues told another scientist about my idea, and luckily, it turned out, that person knew the director of the mechanical factory in Norilsk. So we decided we would write a letter to that director. Several months later, we received a long telegram from the factory, which said: "We are ready to accommodate a group of 20–40 students and employees." The letter said that we could live in trailers and eat at a local canteen. We still didn't know what kind of work we would do there. We knew that it would be mostly physical labor: digging, throwing, carrying, and repairing. I had to find a group of trustworthy people to take with me. I ended up recruiting a lot of my friends and colleagues—people I could trust. The first group had thirty-three men in it. Women wanted to come, too, but we decided to leave them at home, especially since the telegram said they would feed us there. On June 14, 1969, three others and I were the first ones to travel to Norilsk. When we arrived, we went to find the director of the factory with whom we had corresponded. He told us that the director of the Norilsk railroad was looking to hire new people.

Norilsk is an outstanding place, and Russia is extremely lucky to have it. You see, Norilsk has one of the largest deposits of copper, nickel, and platinum in the world. Yet the closest Yenisei River is 95 km away from Norilsk. So they needed to build a railroad to transport natural resources. The Soviet government started building Norilsk in 1935. They used forced labor in construction. Thousands of people died in labor camps there. In fact, the director of the Norilsk railroad used to be in charge of a labor camp and he used Gulag prisoners to build the railroad. When we showed up at his doorsteps in the summer of 1969, he said he could find work for us. We were young and energetic. Some of us were students; others were junior research scientists. In short, they were happy to hire us. Later, that same man also asked me to stop by his house after work. When I came by, he closed the door and started telling me about his life. He needed to pour his heart out, and he thought I would be a good listener. Afterward, I visited him about once a week and we talked. He gave me his perspective, of course. He didn't talk about Gulag prisoners or the construction of Norilsk. Unfortunately, I didn't ask any questions either. It was natural for him to have used forced labor. He saw prisoners as cogs in the machine, which had to play their part in building the city.

Our first trip to Norilsk allowed us to make 1,035 rubles each. We were blown away. Almost everyone wanted to go work in Norilsk after that. After we left, a new group from Chernogolovka arrived right away. About 120 people went to work in Norilsk that first summer.

What exactly did you do in Norilsk?

We repaired the railroad. We had to move rail tracks with our bare hands. It was extremely strenuous physical labor. I remember during our first night, they woke us up at 2 a.m. They needed us to unload a railway car full of crushed stone and distribute it evenly on the rail tracks, before a train arrived. We soon found out that there were no cooks there. So we asked our wives to come. Men would buy groceries and women would make meals for us. We bought deer meet from the locals. We worked so hard we fell off our feet by the end of the day. We didn't know how much they would pay us for our work. We hoped to make 500 rubles at best. No one could have even dreamed of making 1,000 rubles, which was almost as much as the annual salary of a junior research scientist. Most people volunteered because it allowed them to see a new place and spend time with their friends. There were a lot of like-minded people on that first trip. The railroad director wanted us to repair 4 km of rail track. We repaired 8 km instead, so he paid us 1,000 rubles each. Later on, the *shabashka* movement became quite popular in Chernogolovka. When people came back from these trips, they could buy refrigerators, TV sets, and even cars. The movement continued to grow: 120 people participated in 1969, 250 people traveled to Norilsk the following summer, and 600 people went in 1971. Many wives wanted their husbands to go, since they could earn a lot of money there. Dubovitskii didn't like it: he was afraid of work accidents. But thank God, nothing like that happened.

Did you go to Norilsk every summer after that?

Not every summer. In 1971, I decided that we had seen enough of Norilsk. We wanted to travel even further, to the Far East. So we started looking for opportunities to go work in Khabarosvk Krai, Primorsky Krai, Kamchatka, Sakhalin. The destinations of our trips were truly incredible. We built roads, houses, cow-sheds, and irrigation canals. We did so many good, useful things. No one told us to do them. Yet we worked hard, because we had tremendous motivation. We invested our resources, and sometimes it came at the expense of our health. But we all made good money. Of course, we could have survived on our salaries, just like everyone else did. But then we could not have bought TV sets, furniture, and cars. This money helped us get up on our feet financially. Later, we would take our children with us, too, since we could not leave them at home alone. They saw many beautiful places growing up.

You mentioned that Norilsk was built by prisoners. Did you encounter any prisoners while you worked there?

No, I didn't. But we were not interested in such things back then. I could have asked the railroad director about labor camps, but I didn't. I didn't know much about the Gulag at the time. I started paying attention to such things only later. In fact, I began to critically evaluate Soviet life after I met Kronid.

Kronid Arkadievich Liubarskii moved to Chernogolovka in the late 1960s. He lived in an apartment building right across the street from ours. I think it was my neighbor, Isai Braginskii, who introduced me to him. Braginskii had known Kronid from their time in Simferopol and invited him to teach physics and astronomy at a local school. Liubarskii was an astronomer and a candidate of physical and mathematical sciences. But he was also a dissident and a human rights activist. Unlike me, he was keenly aware of the Soviet past and he knew the history of the Gulag quite well.

Does it mean that it was possible to get access to information about the Gulag?

Not really. One had to pay a high price for such access. Kronid paid for it with his own freedom. He spent five years in jail. When he was finally released in 1977, they forced him to leave the country.

I met Kronid in 1968, shortly after he came to live and work in Chernogolovka. I guess I can talk about it now. He gave me samizdat literature to read, including *The Chronicle of Current Events*, Solzhenitsyn's *First Circle* and *Cancer Ward*, Avtorkhanov, and other books. In fact, this is why he was arrested in the first place. They charged him with distributing illegal literature. When the KGB searched his apartment on January 15, 1972, they discovered an extensive samizdat library. Liubarskii was arrested the same day. Kronid's trial took place in the fall of 1972 in Noginsk, a town not far from Chernogolovka. Few people were allowed to attend the trial. I remember Dubovitskii brought some party and Komsomol members from the institute to the trial. They came by bus, and their main purpose was to demonstrate that the scientific community disapproved of Liubarskii's activities.

My colleague, Oleg Efimov, and I were not invited, but we decided to go there anyways to see what was happening with our own eyes. There was a large barn lock on the court building. Next to it, we saw Andrei Sakharov, Elena Bonner,[9] and several other dissidents. We walked past them and said hello, but we didn't stop. We knew that we were being watched. Honestly, I was afraid. I had already been summoned for interrogations at KGB headquarters in Lubianka, and also in their office in Noginsk. They asked me if Liubarskii had given me any samizdat literature to read. I naturally denied everything. I told them we mostly talked about science and a little bit about politics. We also translated a book together. Kronid was a very talented person. He taught physics—many of his former students still remember him warmly. He also was working on a physics textbook. He even gave me some chapters to read. It was a wonderful textbook. I was jealous of students who used it (laughs). After Kronid was arrested, they published this book, but without his name. They simply crossed it out.

This is how we lived back then. This is how we lied. We were afraid to tell the truth. We had no rights or freedoms. Once, I remember I brought up the Soviet invasion of Czechoslovakia in a private setting. Several days later they scolded me at a party meeting for this, and I had to defend myself. Someone must have overheard me and reported on me. In this sense, my relationship with Kronid had a huge impact on me. I learned many new things from him. I learned to be more confident, to be more conscientious, and critically-minded. I learned to question the lofty ideological proclamations and promises of the Soviet regime.

In fact, Kronid and I did translate a book together. He knew English quite well, and Oleg Efimov recommended that he ask me to help him translate *Chemical Evolution*. It was a book by Nobel Prize laureate Melvin Calvin. My English was not that great, but I always wanted to improve it, so I agreed to help Kronid, and we worked on the translation together. The book was published before Kronid was arrested. This helped my case during the interrogations, because I could tell them that we had worked on that book. As I already mentioned, I denied that Liubarskii had given me any samizdat literature.

Rustem Bronislavovich, you said that you met Liubarskii in 1968 and went to Norilsk in 1969. So by then you should have heard something about the Gulag.

You know, the transformation of my views, my political awakening, so to speak, was a long and gradual process. I was always a curious person, but Liubarskii opened my eyes to the realities of political life in the Soviet Union. Nowadays, when some people say that we need a ruler like Stalin, they don't know what they are talking about. When you tell them about the millions of people who perished in the Gulag under Stalin, they respond: "So what? We built a country that survived the Second World War. We modernized and industrialized under Stalin." But these are such different things! You cannot really compare them. These people simplify everything. They don't want to think for themselves. The Soviet system had no respect for the individual. If I am not mistaken, it was Stalin who said that we have "cogs and screws" instead of people. And, of course, we had the Communist Party, which was omniscient. This was one of the main tragedies of life in the Soviet Union. They told us that the party knew what was good for us and that we simply had to accept it, and most people did. It is difficult to learn how to think for yourself, how to evaluate your actions. But it is necessary.

Today I can say that I am glad that I was summoned for interrogation to Lubianka. I was inside that building. I saw the staircase that Solzhenitsyn described in his writings. Today they are talking about restoring the statue of Felix Dzerzhinsky on Lubianka Square in Moscow.[10] But Dzerzhinsky played a leading role in orchestrating the terror. Why do they need to restore his statue? The Soviet system is on its way back. In a way, it is back already. For example, they brought back the old national anthem. Yes, I knew it well and I used to like it. But why did they need to bring it back now? They replaced the lyrics, but as long as my generation is alive, we will always hear the old lyrics: "And Stalin our leader with faith in the people inspired us to build up the land that we love." They beat these words into our heads all our lives. I hear them every time they play the new anthem. It takes a lot of work, first, to realize that you are not free and, second, to start liberating yourself, or, as Chekhov put it, to "squeeze the slave out of yourself, drop by drop." It took me a long time.

Did you know about dissidents or dissident trials before Liubarskii moved to Chernogolovka?

I heard something about the trial of Joseph Brodsky.[11] I knew that he wrote poetry and that he was accused by Soviet authorities of being a parasite who didn't want to work. We didn't know exactly what kind of crime he committed. We just knew that he was exiled to the north. Then there was the Sinyavsky and Daniel trial. We listened to some updates on the Voice of America. There might have been some other trials of dissidents, but I don't remember them now. Soviet newspapers usually didn't mention them at all, and when they did, it was in a highly critical manner. There was no truthful information available about the trials. I remember some people occasionally protesting against the regime, but we could not relate to them. Why would they protest against the regime that provided us with housing and good salaries? What else did they need? Why could they not enjoy their lives like everyone else did? Because overall we thought we had decent lives. Besides, our community was not far from Moscow. If only there had been more buses running between Moscow and Chernogolovka, life would have been prefect. We didn't need much to feel content, and of course, we didn't espouse any revolutionary views. But then I met Kronid.

Do you know how many scientists in Chernogolovka knew Kronid personally?

It could not have been many. Oleg Efimov, his wife, Isai Braginskii, and myself. I shared the literature that he had given me with some of my coworkers at the institute. But, of course, I asked them not to tell anyone about it.

Were there any other KGB searches in Chernogolovka before 1972?

No, it was a quiet place. Even the search of Liubarskii's apartment was rather random. They arrested several dissidents in Moscow and found evidence that pointed to Liubarskii. The KGB didn't expect to find an extensive library at Liubarskii's apartment. They found it by accident. Many Chernogolovka scientists, though, didn't read samizdat. They had their science, and that was enough for them.

What kind of literature did you borrow from Liubarskii?

I borrowed Solzhenitsyn's *Cancer Ward* and *First Circle*, Avtorkhanov's works, and *The Chronicle of Current Events*, which Kronid helped publish.

Did you continue to borrow samizdat from Kronid's wife, Galina Salova, after his arrest?

Rarely, mostly books on how to behave during interrogations.

You mentioned earlier that Kronid opened your eyes to many things. What exactly did you have in mind?

He helped me get rid of fear, at least partially. I was still afraid, of course, but less so. That fear was ingrained in us. Kronid also showed me that even though I graduated from Fiztech, I still knew very little. He showed me that there were many exciting things in the world besides science. He was very well educated. He told me about poetry and prose that I had never even heard of. He even told me about some areas of physics that I knew nothing about. He made me aware of the world beyond the Iron Curtain. How did he have all this information? He was inquisitive, and he read a lot.

You see, in the Soviet Union, they told us what to think and how to behave. People would come to work and wait for instructions from above. Our children are different. They are entrepreneurial, they take initiative, and this is one of the biggest benefits of a market economy. It is only natural that an individual thinks about himself and his family first, rather than his Motherland. When you cannot take care of yourself, you only do minimal work, and expect the state to take care of you. We simply didn't understand how our society functioned. We were, in a sense, helpless.

We now know that the Soviet state often failed to provide its citizens even with basic foodstuffs. Why was this the case?

I can explain this to you. In the USSR, they charged 6 kopeks for 1 kg of potatoes, while its production might have cost 10 or even 15 kopeks. There was a huge disparity between sales prices and real prices. Prices should to be set based on how much money one invests and how much profit they need to make. But this was not the way things worked in the Soviet Union. Gosplan, or the State Planning Committee, set all prices artificially, without taking market considerations into account.[12] Besides, prices should fluctuate across the country, based on different factors, but we had the same price for a product all across the Soviet Union. This was not sustainable. People got used to it, but the economy could not survive like this for a long time. The role of the state is to create conditions that allow people to take initiative. People need to pay taxes, but the state has to implement laws that would allow people to establish and maintain successful businesses. When the state attempts to nationalize businesses, it undermines entrepreneurship and destroys motivation. It makes people much less inclined to take initiative, and it is bad for the economy.

Did you realize that Kronid was a dissident when you borrowed all this literature from him?

I did, although he never accentuated that.

Did you know that he had an extensive samizdat library?

Of course, I saw all the books in his apartment when I came to visit.

Did you ever think about becoming a dissident yourself?

I didn't quite realize what it meant to be a dissident. I wanted to work and to do physics, and I had no reason to become a dissident. Dissidents usually protested against something, but I didn't fully understand against what they protested. The state provided us with housing and decent salaries. Besides, I had no idea how people lived in the West, so I had no point of reference there. I simply didn't know what I would protest against. Would I protest against the Soviet invasion of Czechoslovakia in 1968? I was not ready for that yet. I remember when eight people held a demonstration, protesting the Soviet invasion at Lobnoye Mesto in Moscow. I thought they were extremely brave. But I also felt sorry for them, because it was obvious that they would end up in jail. You should keep in mind that we lived in a bubble, in a sense. First, when I was a student at Fiztech, I had few interests beyond science. After I moved to Chernogolovka, I became part of the scientific community, which was also quite isolated. Nowadays, you can access the Internet and read about anything you want. Back then, however, we knew little. The only information we had was from Soviet newspapers, which condemned dissidents ruthlessly. We didn't know much about the Soviet rulers either. Only after Kronid started giving me samizdat did I learn that a lot of top Soviet officials were poorly educated. They were ignorant people, and yet they named cities after them. But even then I had no personal reason to become a dissident.

I don't think Kronid Arkadievich had any personal reasons to become a dissident either. It seems to me that he simply wanted to have access to uncensored information. In his last word at the trial, he explained that he resorted to samizdat, because the official newspapers didn't tell the truth.

Yes, it was important for Kronid to have access to information, and he had many friends in Moscow dissident circles. I only gained access to uncensored information after I met Kronid. In general, it is important for a scientist to have access to various kinds of information. But it was never enough for Kronid to simply know about recent scientific developments. He also wanted to know the truth about Soviet politics and everyday life. And this came in conflict with the official party ideology, which aimed to impose its own narrow worldview on Soviet citizens.

Could your relationship with Kronid Liubarskii have had negative consequences for your professional career?

Yes, it could have, but Dubovitskii didn't let this happen. The KGB summoned me for interrogation several times, and every time I denied that Kronid had given me any samizdat to read. Eventually, they let me go because I think Dubovitskii interfered on my behalf.

When did they summon you for interrogations?

Kronid was arrested in January 1972, and I was summoned for interrogations in March and April 1972.

Where did the interrogations take place?

The first one took place in an apartment building in Chernogolovka, not far from here. They asked me what I knew about Liubarskii. Someone called me from the First Department of our institute and asked me to come.[13] They didn't explain why, but I had expected it after Liubarskii was arrested. We talked for a long time. Unfortunately, I didn't write down what we talked about. The person I talked to asked me not to mention our conversation to anyone. Of course, I told Oleg Efimov, who said to me that he had also been summoned for interrogations. After that incident, I had to go to Lubianka. It was a day-long interrogation, which lasted from 9 a.m. to 5 p.m. There was also another time, when they told me to come to the KGB office in Noginsk. That time there were two people in the room, and they took turns asking me questions. As I already told you, I denied that Kronid had given me any literature. I was scared. I had small children, and I could have been fired or expelled from the party. I could have lost everything.

Rustem Bronislavovich, when and why did you join the Communist Party in the first place?

I joined the ranks of the party in 1968, before the events in Czechoslovakia. I thought that if we wanted to get things done, we had to be party members, since all decisions had to be approved by the Communist Party. For example, my friends and I wanted to invite artists to perform in Chernogolovka. The local party organization had to approve the invitations. At some point, we also wanted to organize camping trips for institute employees—the party had to give us its approval for that, too.

Did you hold any positions in your institute's party organization?

Not really. In the early 1980s, I was in charge of organizing a system of rationed food parcels (*systema zakazov*), since there were food shortages in Chernogolovka. This is how it worked. First, we got in touch with the October regional committee of the CPSU, and they helped us make arrangements for food provisions. We would then hire a truck and go to a large grocery store on the Leninskii Prospect in Moscow. Later, we would distribute these food parcels among the institute's employees.

But that was hardly the main purpose of the institute's party organization.

True, its main purpose was the ideological education of the institute's employees. The party had to disseminate the ideas of Lenin, Marx, and Engels, making sure that no one diverged from these ideas, either to the left or to the right.

Did party membership take up a lot of your time?

It took me away from my research. Every month I had to attend party meetings, where we voted in favor of various resolutions proposed by the Central Committee. Almost no one challenged those resolutions; besides, we rarely had to vote on truly controversial issues.

You mentioned earlier that it is important for a scientist to have access to information. What sources of information were available to Soviet scientists in the 1960s and 1970s?

It depends. We could use the institute's library to read foreign scientific journals. As far as political and social information goes, we could try to tune in and listen to the Voice of America broadcasts, or read samizdat.

Did you have a radio set at home?

Yes, we bought our first radio set in 1966–1967. It was a small radio set. I remember one time we sat down for dinner and Maksim, our older son, asked: "Papa, why do they broadcast the Voice of America on radio so frequently?" Rimma and I were both speechless. After that we made an effort not to listen to foreign broadcasts, when our children were around. If they had said something like this at school, this could have gotten us in trouble.

Did you also read Soviet newspapers or magazines?

I certainly did. I subscribed to *Pravda*[14] all my life. In fact, it was a requirement for all members of the Communist Party. It was a highly ideological newspaper, but it sometimes published good articles about sports and everyday life. I also subscribed to *Literaturnaya Gazeta*, and from 1966 to 1991, Rimma and I were the subscribers of *Novyi Mir*.[15] Nowadays, I listen to *Echo Moskvy* and read *Novaya Gazeta*.[16] They both strive to tell the truth, although few people want to hear it.

You see, Kronid and other dissidents were not the enemies of Russia. Kronid loved and respected Russia, but he wanted its people to know the truth. He wanted them to be able to speak freely and to read whatever they wanted. He wanted the Soviet regime to follow the Soviet Constitution. Kronid himself was a Soviet citizen. He just knew much more than an average Soviet citizen, and since he knew the facts, he could compare them and conduct a critical analysis of the situation. Kronid was a freer person, too. Unlike him, the rest of us didn't know much about politics, except for the information they told us during political seminars. We focused primarily on science. Work occupied all our time. In this respect, being a scientist is different from many other professions. When you work as a taxi driver or a sales person, you can work for eight hours a day, and then be done. My work doesn't stop when I go home. We simply didn't have time to read as much, or even to start thinking differently about life around us.

Do you remember if Kronid moved to Chernogolovka before or after the events of 1968 in Czechoslovakia?

He came here prior to that.

Did you discuss the Soviet invasion with him?

Yes, we discussed it. He was upset that he didn't know about the demonstration at Lobnoye Mesto. Those people were his friends and I think he wanted to be there with them.

You mentioned earlier that someone overheard you discussing the invasion with some of your colleagues at work and reported on you. What happened exactly?

Several of my colleagues and I discussed the Prague Spring and said that our party should have anticipated the developments in Czechoslovakia and that the invasion could have been avoided. One month later, the institute's party organization received a letter from the October regional committee of the Communist Party. This letter said that some members of the institute's party organization didn't fully appreciate the severity of the situation in Czechoslovakia. The letter mentioned my name, along with some of my colleagues' names. We defended ourselves, of course, saying that we discussed the international situation but didn't criticize the invasion.

Did the party organization also publicly condemn your relationship with Kronid Liubarskii after the Liubarskii Affair of 1972?

No, it was different that time. After Kronid was arrested, I was instructed not to mention my friendship with him to anyone. In fact, they barely mentioned the Liubarskii Affair at party meetings at all. The official line was that, yes, Liubarskii lived in Chernogolovka for some time, but he didn't work at local research institutes and thus had no relationship with the local scientific community.

Rustem Bronislavovich, changing the subject, what do you remember about Stalin's death?

My family still lived in Lviv, and I was in ninth or tenth grade. On the day we learned that Stalin died, our neighbor stopped by our apartment and told us the news. I remember my mom saying "Yes, our idol is dead." She said it somewhat sarcastically, but then she added: "It's a pity. What are we going to do now?" My mom's words stuck in my head. I also remember there were public meetings in the street. People made speeches in the parks. They took turns praising Stalin. I already told you that our parents didn't tell us anything about the 1930s. Nor did we learn about the repressions in school. I would say, in general, people loved Stalin. Now I understand that the October 1917 Revolution brought a lot of harm to Russia and its people. If it was not for the revolution, we would live very differently now.

I am a little surprised to hear this. Many of your colleagues would argue that the Russian Revolution had a positive impact of Russia's development, contributing especially to the development of science.

The Revolution forced many Russian intellectuals to leave the country. Among them were people like Sikorsky, who emigrated to the United States, and became a pioneer of aircraft construction, and Zworykin, a pioneer of television technology. Many talented scientists fled Russia after 1917. Kapitsa left, too, and he worked in England until he was prohibited from leaving the USSR during one of his visits there. Landau was almost executed in the 1930s. So those scientists who claim that the Revolution had a positive impact on the development of Soviet science simply lived under ideal conditions, largely thanks to Dubovitskii and Semenov. Many conducted research on military-oriented topics, which had tremendous financial support of the Soviet government. Yet, no one knew about their research outside the Soviet Union.

And what do you remember about Khrushchev's speech at the Twentieth Congress of the Communist Party in February 1956?

I was a high school senior. They didn't read Khrushchev's speech in schools, but I knew approximately what it was about. I also remember hearing about the political struggle and the release of Gulag prisoners. The latter development was outstanding. Then I became a student at Fiztech in 1957, and I hardly had any time left to care about politics. You see, we were all raised to think the same way. We all sang similar songs, in which we praised Lenin and our Soviet Motherland. We all became Octobrists[17] and had to wear little pins with a picture of young Lenin with curly hair. He looked so cute and innocent in that picture. For a long time, Lenin was almost like a saint to us. Then in the early 1960s, when they told us that we would build Communism in twenty years, we believed it. In general, I found socialist ideas attractive, although I realized that a lot of them were utopian. What does it mean "to each according to his needs"? People have different needs. In the Soviet Union, people used to come to work and wait for someone to tell them what to do. There were no incentives to work hard.

Yet many scientists stayed loyal to the Soviet system.

Yes, but they didn't care about politics that much. The state made it possible for us to get a good education. So we thought that we owed what we had to the state. This is simply the way we were raised.

And what did you think about the one-party political system?

Few people took elections seriously. Most people voted because if you arrived at the polls early enough, around 6 a.m., you could buy some food, clothes, or even books, which they brought to the polls to increase voter participation. Besides, we didn't really know anything about the West or their political system. We knew that they had capitalism over there, and that under capitalism 10 percent of people were extremely rich, while everyone else was extremely poor. They showed us photographs, in which poor people in capitalist countries were going through garbage looking for food. They also made a concerted effort to suppress all foreign broadcasting. Soviet newspapers were the only source of information available to people in the USSR. So we spent most of our time working for set salaries. In the summer, we went mountain climbing and hiking. Our trips to Norilsk allowed us to step outside of our regulated existence for a while and earn some real money.

Did you think that the Soviet system needed reforms?

Well, you see, we were raised to think that there existed only two ideologies in the world: capitalism and socialism. Under capitalism, they explained to us, only a small percentage of people were wealthy, while the rest lived in poverty. They told us that in comparison to the vast majority of people living in capitalist countries, we were well-off. Since we were not allowed to travel abroad, we didn't know if that was true. I remember in the 1950s, when we had ration cards, I dreamed that one day I would be able to eat as much bread as I wanted to. So when they finally abolished ration cards, we considered it to be a sign of progress. We thought that people in the West lived in much worse conditions than we did. It is scary when a government deceives its own people like this. Today it would be impossible to do that—the Internet makes it impossible to conceal information on such a large scale. But back then the authorities did as they pleased. There were very few individuals, courageous individuals, who tried to make people aware of what was going on and open their minds a bit. Kronid was one of them. This is why I always had sympathy for Kronid and other dissidents.

Rustem Bronislavovich, did you personally think that the Soviet system needed reforms?

Yes, of course I did. First, they had to stop lying to us and start telling us the truth. There was a lie at the very foundation of Soviet ideology: "From each according to his ability, to each according to his needs." I realized that it was a utopian slogan, which went against human nature. It is only natural for an individual to think about himself and his family first. Only when you and your family are doing well can you start thinking about sharing with other people. But as I already mentioned, I didn't even think about such things before I met Kronid. Prior to my friendship with him, I took many official party slogans and proclamations for granted. As a high school and college student, for example, I was told that our country was recovering from a horrendous war and that we had to invest in heavy industry, even if it came at the expense of all other industries. Then, they said, we had to be strong enough to defend ourselves against the capitalist countries that had surrounded us with their military bases and threatened to destroy us. In the 1960s, they told us that we had already built socialism and that in twenty years our children would live under communism. We were completely brainwashed. They said we just had to work harder and everything would be fine. Even if someone had doubts, they would never voice them publicly. People like Kronid changed all that. Kronid helped me see the world differently. When I tried to challenge him sometimes, defending the Soviet system, he defeated all my arguments without much trouble.

Unfortunately, our friendship didn't last long. He was arrested in 1972 and sentenced to five years in the Gulag. Kronid helped me develop my own worldview, which was different from official Soviet propaganda. Yet even after he was arrested, I didn't leave the Communist Party. I saw my membership in the party as an opportunity to get things done and help people. I helped organize cultural events and a system of rationed food parcels, as well as our trips to Norilsk.

Rustem Bronislavovich, when did you first travel abroad as an adult?

Well, as a child I lived in Poland for two years, from 1945 to 1947. After that, I was not allowed to travel until 1979, when I went to Hungary. They simply didn't allow me to leave the Soviet Union. In 1975 or 1976 I applied to go skiing in Czechoslovakia, but they rejected my application. Then in the late 1970s, my colleague and I were invited to come to work at a scientific lab in Hungary. At the time, we worked on trying to synthesize organic superconductors, and a Hungarian colleague invited us to come and visit his lab. It took us almost two years to get an "exit" visa,[18] even though the trip itself only lasted for one week. My next trip abroad was in 1987 to Poznan in Poland. Rimma had attended a conference in Tashkent prior to that and she met a scientist from the Institute of Molecular Physics. He had his own lab in Poznan and he first invited Rimma, then me. In 1988, I traveled to Yugoslavia, and in 1990, I also went to Tübingen in Germany to attend an international conference there. It was shortly after the fall of the Berlin Wall. In the 1990s, I went to conferences in Portugal, the United States, and other countries.

Why did they not let you travel prior to 1979?

They didn't explain, although later someone told me that it was because of my involvement in the Liubarskii Affair. Traveling was not a priority for me anyways.

Did you not want to travel and see the world?

I certainly did, but how could I afford it? My salary was only 105–130 rubles. When I defended my candidate of sciences degree in 1972, my salary became 175 rubles. Rimma made about the same. It was barely enough to support our family.

What did you think about your first trip abroad?

It opened the world to me. I realized that we would not have such abundance for a long time. I remember we walked into a large grocery store in Budapest and were stunned to see such a variety of goods and produce. Budapest was a beautiful city, too. It took us two years to get an "exit" visa, as I already mentioned. The process was extremely complicated. I had to prepare a large dossier, including an evaluation of my scientific activity and my moral standing. It is hard to imagine, but it had thirteen signatures on it. Thirteen different people had to sign this piece of paper, so that I could travel abroad—first, a secretary of our department's party organization, a head of the trade union and the Komsomol organization had to sign it; then, secretaries of the institute's party organization and the institute's director; finally, members of the Academy of Sciences in Moscow had to approve all these signatures. Most of these people had never seen me before and had no idea who I was, but they put down their signatures anyways.

Did you think that contacts with foreign colleagues were important for your research?

We simply didn't understand whether they were important or not. We didn't realize that we lived in an isolated environment, separated from the rest of the world. We didn't know, for example, that our foreign colleagues had much better equipment. Only when we started traveling abroad did we find out that our equipment was outdated. In this respect, Soviet science lagged behind significantly.

And who paid for your trips abroad?

Usually the Academy of Sciences did. They also issued us some amount of American dollars, telling us to exchange them into a local currency. When we went to Hungary, they said we could only spend this money in case of emergency and had to return the rest. Of course, we ate mostly at the conference and didn't even consider going to a restaurant. I remember one time this put me in an awkward position. I met outside of the conference with two of my colleagues—a Portuguese man and a French woman—to discuss our research. My French colleague kept suggesting that we continue our conversation at a restaurant. But all I could think about was: "What restaurant? How will I pay for it?" So I insisted that we stayed outside, pointing out that it was a beautiful night, and we should enjoy it. Before leaving, I exchanged the money back to dollars but lost on the exchange rate. I was terrified by the prospect of explaining what happened at the Administration of External Relations at the Academy (laughs). As soon as I came back, I returned all the money I had left after the trip; luckily they didn't ask any questions. This is just the way we lived back then. It is not surprising that nowadays many Russian scientists choose to emigrate and go work abroad. Several of them recently got a Nobel Prize. It is only natural that they leave: scientific research cannot flourish amidst corruption. There are many talented Russian people, whose skills and talents are appreciated in the West. In fact, my son and his family live in the United States. Yet I never wanted to leave Chernogolovka. This is my home.

Rustem Bronislavovich, how did you view Gorbachev's coming to power in 1985?

Extremely positively. All previous leaders. . . . Brezhnev could not even pronounce words by the end of his life. After Brezhnev's death, Andropov started introducing more repressive policies. Chernenko, who came after Andropov, could barely speak. By contrast, Gorbachev spoke freely, without any notes. He argued that we needed socialism with a human face, which would prioritize people's well-being. We were dumbfounded. This is why we welcomed Gorbachev's rise to power. He also brought many new talented people with him, including Boris Yeltsin, who was a secretary of the Moscow city party committee. They were both young, full of energy. You see, no one expected that the reforms would lead to the collapse of the Soviet Union.

Nowadays many people, including some of your colleagues, criticize Gorbachev. They say that he should have started with economic, not political reforms. Some point to China and the reforms of Deng Xiaoping as a model. Do you agree with such assessment?

Gorbachev tried to carry out economic and political reforms at the same time. But he ran into insurmountable economic problems. The Soviet economy was completely cut off from the rest of the world. The only thing we sold was natural resources—oil and gas—and when oil and gas prices dropped, the Soviet economy began to crumble. We invested most of our money into the military and the army. In the 1980s and 1990s, our country's economy kept deteriorating. So Gorbachev didn't neglect the economy, but he didn't fully know what he wanted to accomplish. What exactly was it, "socialism with a human face"?

It was obvious that Gorbachev didn't even consider trading socialism for capitalism. But he did recognize the importance of individual initiative. This is something that I respect about his reforms to this day. Besides, he was young, optimistic, and he wanted to improve things. Of course, he ran into tremendous opposition. He made mistakes, too. But at the end of the day, he changed the world. He saved the world from another war. He wanted to achieve peace, to reduce arms, and to reestablish relationships with European countries. How could I not support these goals? We lived in the world, divided into two camps. In the 1970s, they kept updating us on how many times we could destroy each other—five times over, ten times over. Why did we need so many weapons? At the same time, our productivity was much lower than in the United States. So we achieved military parity, but our productivity lagged behind significantly. How was this possible? Well, ordinary people had to pay the bill for that military parity. It was achieved at their expense, and as a result they could not even buy basic produce in Soviet stores.

And what did you think about Yeltsin?

I respected him, but I didn't want him to be my president. He was a drunk and didn't keep his promises. I wanted to have an intelligent, educated, and knowledgeable person in charge. But there was no alternative. Besides, Sakharov, Sobchak, and Afanasiev supported Yeltsin. I also respected Yeltsin for his courage during the August putsch of 1991.[19] It is important to keep in mind that Yeltsin was President of Russia during a very difficult time, too. In the 1990s, there was a lot of crime. Some people became incredibly rich in a day—the so-called oligarchs like Berezovsky, Abramovich. There was a threat of civil war.

Did the Soviet collapse take you by surprise?

It took everyone by surprise. The question, though, is why it fell apart. We lived in a dictatorship, where violence was key to holding the country together. In 1991 no one was there to defend the Soviet Union. In fact, different Soviet republics rushed to leave the Union. This means that the country was created by force.

When the Soviet Union collapsed, at first we were terrified. We were used to the party making all decisions for us. But now that the party was gone, we had to start thinking for ourselves. It was extremely hard. In fact, I experienced it firsthand in the early 1990s, when I was elected as a member of the Council of Deputies in Chernogolovka. We quickly realized that we were completely helpless without state funding. We wanted to help people, but we barely had any resources for that.

NOTES

1. *Moskal* is an ethnic slur for ethnic Russians, usually used by Ukrainians, Belarusians, and Poles.

2. Lviv is a city in western Ukraine. From 1939 to 1991, it was part of the Soviet Union. Here Liubovskii uses "Russia" and "the Soviet Union" interchangeably.

3. Liubovskii is mixing up dates here. The Soviet invasion of Czechoslovakia took place in August 1968.

4. I. V. Kurchatov Institute of Atomic Energy was originally established in 1943 as the "No. 2 Laboratory of the USSR Academy of Sciences." Its main purpose was to develop nuclear weapons and design nuclear reactors. Beginning 1955, the institute also started working on plasma physics and thermonuclear fusion. The Institute was named after its founder and first director, a Soviet nuclear physicist, Igor Kurchatov.

5. *Sharashka* is a Russian slang expression, meaning a poorly run organization. In the Soviet era, *sharashka* was an informal name for research laboratories in the Soviet labor camp system, the Gulag. For more information on *sharashkas*, see Asif Siddiqi, "Scientists and Specialists in the Gulag: Life and Death in Stalin's *Sharashka*," in *Kritika: Explorations of Russian and Eurasian History*, Vol. 16, No. 3, Summer 2015, pp. 557–88.

6. Yegor Gaidar was the prime minister of Russia under Boris Yeltsin, in 1992. He carried out a series of economic reforms, known as "shock therapy," which aimed to transition Russia to a market economy.

7. Liubovskii is referring to the widespread shortages of consumer goods in the late Soviet era.

8. *Subbotnik* was a voluntary community work day.

9. Elena Bonner was a wife of Andrei Sakharov, a prominent Soviet scientist and human rights activist. Both played an important role in the Soviet dissident movement.

10. Felix Dzerzhinsky was the founder and first head of the Soviet secret police. His statue on Lubianka Square in Moscow was toppled shortly after the collapse of communism in the Soviet Union.

11. Joseph Brodsky was a Russian and American poet and the winner of the 1987 Nobel Prize in Literature. In 1964, he was put on trial and sentenced to five years in a labor camp for "social parasitism." Soviet authorities claimed that his poetry didn't constitute work.

12. Gosplan was the state planning agency set up in 1921. It was in charge of central economic planning in the USSR.

13. The First Department was a part of the KGB and was responsible for political security and secrecy.

14. *Pravda* was the official newspaper of the Communist Party of the Soviet Union.

15. *Novyi Mir* (New World) was a highly influential literary journal in the Soviet Union. It published texts by writers such as Aleksandr Solzhenitsyn and Boris Pasternak. For an excellent study on the journal and its influence on Soviet citizens, see Denis Kozlov, *The Readers of* Novyi Mir: *Coming to Terms with the Stalinist Past* (Cambridge: Harvard University Press, 2013).

16. *Echo Moskvy* is a commercial radio station, which offers news and a variety of talk shows on social and political issues, representing different points of view. It is currently one of the few independent media outlets in Russia, which provides a public platform to critics of the Putin regime. *Novaya Gazeta* is a Russian newspaper, famous for its investigative journalism and critical coverage of Russian political and social issues.

17. The Octobrist organization was a Soviet youth organization. Established in 1924, it prepared children from seven to nine years old for joining the Pioneer movement.

18. Soviet citizens needed to obtain an "exit" visa to leave the country even for a short period of time.

19. Liubovskii is referring to the coup of August 19–21, 1991. The coup was staged by the vice president of the USSR and the head of the KGB, who formed the State Committee on the State of Emergency (GKChP), announcing that they had taken over presidential powers from Mikhail Gorbachev. During the coup, Boris Yeltsin went to the White House, the seat of the Russian government, and denounced the Emergency Committee as unconstitutional. The coup failed within two days.

Chapter 6

"Individuals, Not the Party or Society, Should Be Responsible for Their Own Education"

Oleg Nikolaevich Efimov

Oleg Nikolaevich Efimov (b. 1936) was another Chernogolovka scientist closely involved in the Liubarskii Affair. His friendship with the dissident, Kronid Liubarskii, dated back to the early 1950s, when both were members of the Moscow Planetarium. They lost touch for over a decade, but reconnected after Liubarskii's family moved to Chernogolovka in 1967. Until Liubarskii's arrest in 1972, Efimov was a frequent guest at the dissident's apartment. They talked about science and politics. Most important, though, were the samizdat books that Oleg Nikolaevich borrowed from Liubarskii's library. In the interview, Efimov contends that it was not his friendship with the dissident, but the numerous underground books that he read, which opened his eyes to the repressive side of the Soviet system. This corresponded with Efimov's longtime belief that individuals should strive to continuously educate themselves, instead of relying exclusively on the information disseminated by the state propaganda machine.

This chapter sheds light on the doublethink widespread among the critically minded part of the Soviet scientific intelligentsia. On the one hand, they sought access to uncensored information and were some of the most enlightened Soviet citizens. Efimov, for example, took pride in his open-mindedness and critical thinking. He was not afraid of being friends with the dissident or reading samizdat. He also readily acknowledged the economic failings of the Soviet system. On the other hand, even critically minded Soviet scientists continued to support the Soviet state, which invested heavily in scientific development and made scientists elite members of Soviet society.

Efimov's passion for learning and critical thinking is remarkable for an individual living under a highly authoritarian dictatorship. At first glance, Efimov's story was typical for a postwar Soviet youth. Born in 1936 in

Moscow, Oleg Nikolaevich was only five years old, when the Great Patriotic War broke out. The war overshadowed his childhood. First, he lost his father on the frontlines. Then his family had to evacuate to the east. His family suffered so badly during the evacuation in Uralsk that they returned to Moscow as soon as possible. Efimov's mother had to work long hours to support her children and elderly mother. Like many of his peers who grew up during the first decade of the Cold War, Oleg Nikolaevich was captivated with science at an early age. After finishing high school in 1954, he enrolled in the Mendeleev Institute of Chemical Technology in Moscow, becoming a student of the Physical and Chemical Department. Upon graduation, in 1962, he accepted a research position at the Branch of the Institute of Chemical Physics in Chernogolovka. He earned his candidate of sciences degree several years later, and in 1983 he established his own laboratory of electrochemical dynamics at the ICP Branch.

At the same time, Oleg Nikolaevich was never fully content with what official Soviet education system had to offer. In 1948, at the age of twelve, he joined the Moscow Planetarium, which hosted various lectures and study groups. His teachers aspired to educate, not indoctrinate, students. They showed young Efimov and his peers how to conduct scientific research and analysis. They also taught them that learning was an ongoing process and a lifelong responsibility of each individual. In the late 1940s and early 1950s, the Planetarium offered Efimov a welcome respite from the suffocating atmosphere of late Stalinism. Growing up, Oleg Nikolaevich also benefited from his friendship with Georgii Kaminskii, who was one of his friends' grandfather and a polar explorer, and was unjustly persecuted in the 1930s. Kaminskii never talked about his experiences in the Gulag, but he spent hours talking to Efimov about the importance of self-reliance and self-education. When Soviet society started to undergo its tumultuous transformation after Stalin's death in 1953, Oleg Nikolaevich was more prepared for it than many of his peers.

In the interview, Efimov reflects on the gradual transformation of his generation's ideological commitment to the Soviet project. He talks about the stampede at Stalin's funeral and the arrest of Lavrentii Beria in 1953, both of which undermined his trust in Soviet leadership. Khrushchev's "secret speech" at the Twentieth Party Congress in February 1956 left many young people like Efimov with more questions than answers. A lot of them wanted to know why the new collective leadership had remained silent during the mass repressions of the 1930s. "I felt empty, as I was leaving the meeting where they read the speech," Efimov admits. In the 1960s, he read avidly, including Aleksandr Solzhenitsyn's One Day in the Life of Ivan Denisovich *and numerous publications in* Novyi Mir. *After the crackdown of the mid-late 1960s, Oleg Nikolaevich turned to reading underground literature. His friendship with Kronid Liubarskii gave him access to an extensive samizdat library, and allowed him to discuss some of his doubts about the Soviet development with his dissident friend.*

Liubarskii's arrest in January 1972 brought Oleg Nikolaevich face to face with the repressive Soviet apparatus. The KGB summoned him for a series of interrogations throughout the spring of 1972, demanding to know if he had borrowed samizdat books from Kronid Liubarskii. Efimov refused to admit that he had, trying to help out his friend. In late October 1972, he went to the courthouse in Noginsk, where the trial of Liubarskii was staged. It was risky, since the KGB was watching him closely already. But he needed to see what was going on with his own eyes. Despite the dangers it posed, Oleg Nikolaevich maintained his friendship with Liubarskii and his family even after the dissident was sentenced to five years in a labor camp.

Like many other critically minded scientists, Efimov enthusiastically supported Gorbachev's perestroika in the late 1980s. Yet he found himself deeply disillusioned with its outcomes. He resented the drastic decline of state funding of science after the Soviet collapse, which made him somewhat nostalgic for the Soviet era. In the interview, Oleg Nikolaevich struggles to reconcile scientists' newly acquired freedom with the sudden loss of their privileged status in post-Soviet Russia.

Oleg Nikolaevich, let us start our conversation with your childhood. When and where were you born, and what was your family's background?

I was born before the war, on April 9, 1936. I was born in Moscow, even though when I was little I spent a lot of time with my grandparents who lived in Sychyovka, a town in the Smolensk region. My father was a military man. He was killed at the frontlines in early 1942. During the war, my family was forced to evacuate. After we returned to Moscow from the evacuation in 1943, my mom had to work extremely hard to support my brother and me. As a result, we were mostly left to ourselves. Starting in sixth grade, I began to read a lot, especially popular science. This is how my interest in science first emerged. I borrowed most of the books from a local library. I went back to school in 1944. Then in 1948 I became a member of the Moscow Planetarium,[1] which had a tremendous impact on my intellectual development.

I joined the Planetarium largely by accident, but it made a strong impression on me. There were over a hundred people attending various study groups at the Planetarium. They divided us into age groups: middle school, high school, and university students. We spent a lot of time with well-educated, intelligent people, who gave lectures, and also taught us how to conduct research. We did our first scientific experiments there, using a small telescope. It was extremely exciting. When I was in ninth grade, my friends and I published our first scientific paper. There was a special atmosphere at the Planetarium, which focused on learning and self-education. I did well in school, but it was the Moscow Planetarium that shaped my worldview. I spent most of my free time there. I also met many wonderful people at the Planetarium, including girls who were slightly older than us and had a positive influence on our upbringing. One of them was Galina Il'inichna Salova, a future wife of the Soviet dissident, Kronid Arkadievich Liubarskii.

What role did your family and school play in your education?

My school didn't play any significant role. As far as my family goes, I was raised by my mother and my grandmother. But my mom had to work long hours and we barely saw her at all.

What did she do?

She was a director of the children's dining hall. They valued her highly there. Later she became a head of the dining hall at the Moscow Architectural Institute. My mom laid the moral foundations of my upbringing, and to this day, I look up to her. Yet my interest in science and my passion for learning came from the Moscow Planetarium. When I was in high school, I started thinking about where I should apply for university to continue my education. My friends at the Planetarium encouraged me to apply to the Mechanical and Mathematical Department of Moscow State University, where I could keep studying astronomy. But I wanted to study nuclear physics instead, so my plan was to apply to the Moscow Physical-Technical Institute (Fiztech), which offered a cutting-edge education that centered mostly on physics.

What kind of education did your parents have?

They both had technical degrees. My father got a military technical education, and my mom went to a culinary school. Later, she also graduated from the Plekhanov Institute. My mom never interfered with my studies. I earned good grades in school, and my teachers praised me. So she didn't have to worry about me.

And what was your parents' background?

I am not even sure. Unfortunately, I don't know much about my paternal grandparents. As far as my mother's side of the family is concerned, I think her father was a serviceman and an artist before the Revolution. My maternal grandmother only had two years of formal education. Yet she was an outstanding person. Before the war, my grandparents lived in Sychyovka. My father managed to evacuate us several weeks before the Germans occupied the city. If we had not left, they would have shot us, since we were relatives of Communists.

Were your parents Communists?

Yes, of course, my parents were Communists. They were both members of the Communist Party. At the time, it was a necessity. My father was a military man, and everyone who served in the army had to be a party member. My mom was a director of a dining hall. She had to join the party, because it was a high enough position. Being a party member was common back then. But my mom was also a religious person: she was an Orthodox Christian. She had to be discreet about her religious beliefs, though, at least until Gorbachev came to power. Both of my parents were party members, but they were not passionate Communist believers. I didn't know much about their political views, of course. But I can tell you the following story. When I was a little boy, I thought that Stalin was a great leader. One day I brought his portrait home and I hung it up on the wall in our room. It was after my father's death and after we returned from the evacuation. My mom didn't say anything, but several days later she took the portrait down. So, no, she was not an ardent Communist believer. We were an ordinary Soviet family. Like many people at the time, we lived in fear. My grandmother always warned me to be quiet and not to discuss politics with anyone. I didn't really know what she meant. My mom and my grandmother lived in an era when, despite all the ideological excesses, there were still many good, decent people. And both my mom and my grandmother were good, decent people.

Did the purges of the military of 1937–1938 affect your father?

Even if they did, we didn't discuss such things at home. Some of my school friends' relatives had been repressed, though. Let me try to remember their names. One of them was Georgii Dem'ianovich Kaminskii, a grandfather of one of my friends. He was a polar explorer who contributed to studying the Arctic route. He was a remarkable man and an honest Communist. When I met him, he was in his sixties or so. I am not sure whether it was my passion for science or something else, but he enjoyed talking to me. From our conversations, I found out that he had spent three or four years in a labor camp. Back then, many people were imprisoned on false charges, and someone had denounced Kaminskii as well. He was accused of having had connections with foreigners, members of the British intelligence to be precise. They beat him in jail, broke several of his ribs, in order to make him confess. Eventually, he told them that he knew one foreign agent, whose name was Vasco da Gama. The thoughtless interrogator wrote it down. But several years later someone was reviewing Kaminskii's dossier and they found his confession to be amusing. So they let him go. He lived in poverty, but he was an extremely intelligent and educated person, who spoke several languages. I learned so much from my interactions with him. He taught me, for example, that an individual is responsible for his or her own education. Individuals, not the party or society, should be responsible for their own education. How, you might ask? By reading and thinking. I found this advice to be extremely helpful throughout my life.

Did Kaminskii become disillusioned after he was released from the Gulag?

Not at all. By the time he was imprisoned, he had developed his own set of beliefs. His arrest didn't change them, and he remained a devout Communist.

What did he think about his imprisonment?

We didn't really talk about it. Why would he tell a teenage boy about it? He taught me, instead, the importance of critical thinking and self-education. Kaminskii played a central role in my upbringing, which I could only fully appreciate when time came to apply for a membership in the Komsomol.[2] I categorically refused to apply. Everyone scolded me, of course: I was a straight "A" student and I was expected to join the Komsomol. They kept telling me that I had to apply, because then there would be good people in the Komsomol. But I could not understand how my membership would make Komsomol a better organization. By contrast, I was afraid that it would corrupt me. As I mentioned, Kaminskii taught me that an individual should be responsible for his or her own upbringing. Jack London was my favorite writer when I was growing up. I looked up to his protagonists as my role models.

Did the repressions affect any members of your immediate family?

Yes, my maternal grandfather. He was a talented person and an artist. During the war, he ended up living in occupied territory and at some point, he had to work at a German office. When the Red Army approached, he started helping them. He would cross the frontlines and bring information about the location of the German troops. After the Germans left, someone denounced my grandfather and he was sent to a labor camp for five years. After the Gulag, he returned to Moscow. He claimed that he had "lost" his passport, like many other people did at the time. His new passport didn't have any incriminating notes in it. He was a smart man, my grandfather. He didn't like talking about any of this. He never said anything critical about the Communists either. Yet sometimes when I would praise the dekulakization or share something I read in my textbooks, my grandfather would simply say: "Yes, but the kulaks were the most hardworking farmers." He was a religious person, too.

Where did your family evacuate to during the war?

First, we went to Saratov on the Volga River, which was not that bad. Then we decided to relocate to Uralsk. It was supposed to be a more civilized city, but we lived in utterly horrible conditions there. There were bedbugs, lice, and flooding. We didn't stay there long. In 1943 we decided to go back to Moscow. Yet upon our arrival we discovered that our apartment had been occupied by another family. We had some relatives in Moscow, and they let us stay with them. We lived somewhere in the basement, in terrible conditions. There was nothing to eat, except potato peels, which we grinded to make pancakes. But we managed to survive somehow.

Some scientists I talked to said that the war defined their future choice of profession. Was it true for you, too?

No, but I do remember a massive surge of patriotic feelings during the war. We didn't know anything about the repressions back then, so most people were still loyal Communists. I already told you about my friendship with

Kaminskii. Overall, though, we felt patriotic. Later, when I became a member of the Moscow Planetarium, I stopped worrying about political issues altogether. Since the Planetarium was supposed to be a stronghold of atheism, they would come up sometimes. However, our teachers at the Planetarium were dedicated to educating students, not indoctrinating them. So they would tell us about Nicolaus Copernicus, Galileo Galilei, and Giordano Bruno, but they allowed us to draw our own conclusions. They wanted us to become highly educated people, maybe scientists, and they didn't care as much about whether we would also be Communist believers.

Who were your teachers at the Planetarium?

There were many wonderful people. One of them was Felix Ziegel, a brilliant lecturer and a remarkable person. Another one was Victor Bazykin, who mostly directed study groups, but also gave occasional lectures. He was a war veteran and an incredibly charming man, whom we adored. When we were in eighth grade, they started taking us on lecture tours around the Moscow region. We assisted them during lectures, showing slides and instructing the audience about using a telescope. They paid us for our work, too, which was substantial for us children. I saved this money and eventually bought a radio set and a photo camera. Besides, spending time with the lecturers was extremely valuable.

I remember one of the lectures took place in Orekhovo-Zuevo, not far from Moscow. It was in 1953, shortly after Stalin's death. Our tour bus stopped in the park, and I left my fellow travelers to get some water. It was then that I heard an announcement on the radio about the arrest of Lavrentii Beria.[3] They called him a "swine" and an "enemy of the people." I could not believe my ears. I ran back to the bus to share this news with the group, but they refused to believe me at first. This is when our worldview started to crack for the first time. We used to think that Beria was a great man. But then, all of a sudden, they told us that he was an "enemy of the people." There was one more thing, too. When Stalin died, I was in ninth grade. They displayed his body in the Hall of Columns in the House of the Soviets, not far from where we lived. I was one of those mad boys who tried to reach the Hall of Columns by walking on roofs. It was early March, and the pavement was covered with ice. There was a huge pole right in the middle. It was crowded, too, and many people died in that stampede. I remember I saved a boy who was being trampled down. I climbed up on a truck and pulled him out of the crowd. It was a depressing view, which got stuck in my memory.

Did you reach the Hall of Columns?

Of course I did, although it took me several days to get there. As I walked by Stalin, I saw a dead person: his face was dark and covered in smallpox scars. His huge swollen hand was resting on top of his body. People were walking past him. I don't remember anyone crying or wailing.

What made you go there?

My boyish curiosity. At that time, I thought that something big was happening, and I wanted to see it with my own eyes. You see, they broke me down eventually, making me join the Komsomol. They put pressure on me, saying that it was my duty to become a Komsomol member. They used the same arguments later on, trying to convince me to join the ranks of the party, but I refused.

When did you finish school?

I went to school in 1944, when I was eight years old. It was a bit late. I finished school in 1954. Since I was a straight "A" student, I received a gold medal. But as I already mentioned, school was not as important to me. I invested all my time and energy into the Moscow Planetarium.

Did you apply to Fiztech, as you were planning to?

No, I enrolled at the Mendeleev Institute of Chemical Technology instead. There is a story behind it. I was invited to my friend's birthday party, and his uncle was the head of the Physical and Chemical Department at the Mendeleev Institute. We started talking and my friend's uncle asked me what I wanted to do in life. I told him that I wanted to work on a big nuclear bomb (laughs). He looked at me, brightened up and said: "Then you need to apply to my department!" At the time, I was obsessed with nuclear energy and read numerous books about it. Nuclear science was on the rise, and I wanted to be a part of it. This is why I decided to apply to the Mendeleev Institute. Since I had a gold medal, I didn't need to pass any entrance exams, only an interview. However, it was shortly after the Doctors' Plot,[4] and you can probably hear that I cannot pronounce the letter "r" well, which is typical of Jewish people. So when I entered the room and said "*Zdgavstvuite*!" instead of "*Zdravstvuite*!" (Hello!), they looked at me with suspicion. They even checked the nationality field in my application form to make sure I was Russian, not Jewish. It was a comical situation.

But didn't they announce that the Doctors' Plot was fabricated back in 1953?

Yes, they did. But it didn't matter. Anti-Semitism was part of everyday life in the USSR. Besides, there were certain restrictions when it came to higher education. For example, it was almost impossible for a Jew to enroll at Fiztech. There were only three Jewish students studying at our department at the Mendeleev Institute. Yet I met some remarkable people during my studies there. Some of them went on to become prominent people. For example, Valery Legasov played a major role in dealing with the consequences of the Chernobyl disaster. He behaved heroically after Chernobyl happened, but he committed suicide later in life. Other students in my cohort became members or corresponding members[5] of the Soviet Academy of Sciences. As far as I am concerned, I have had two main passions in life: science and alpinism. I started doing scientific research as soon as I enrolled at the Mendeleev Institute. I fell in love with alpinism during my student years, too, and I remained an avid alpinist for the next thirty-five years. As a student, I joined the Sports Club of the Academy of Sciences, which had a strong alpinist division. I could not be happier, partially because I met many incredible people there, including Igor Evgenievich Tamm.[6] I became good friends with his son, Evgenii Tamm, who organized the first Soviet expedition to Mount Everest. Later, I also participated in his expedition to the Pamir Mountains in Central Asia.

What made you fall in love with alpinism?

Well, first, it was a romantic sport, which is hard to compare with anything else. It also helped me form many life-long friendships. I was lucky to have met many wonderful people during these expeditions. Besides, I started coaching amateur alpinists. We traveled all over the country, visiting the Pamir Mountains, the Tian Shan. . . . During one of these expeditions, I met Igor Shchegolev, who later became my colleague at the Branch of the Institute of Chemical Physics in Chernogolovka. He played the guitar marvelously, and he was a witty and extremely intelligent man.

Did you discuss any political or social issues during your expeditions?

Yes, of course, we did. It was a curious time, when many thoughtful, intelligent people started reading Lenin thoroughly, reflecting on what he wrote. And Lenin said that under socialism the economy would achieve much higher rates of production that under capitalism. Yet it was becoming obvious to us that we had been losing the economic competition with capitalist countries. Our economy was stagnating, and it was unclear what we could do to fix it. Then there was Brezhnev, who liked to decorate himself with lots of medals.

You are talking about the 1970s and 1980s now. What about the previous decades?

Well, before Brezhnev we had a "peasant tsar," who also made a strange impression. . . . Khrushchev is remembered for carrying out a revolution, the so-called Thaw. It was a period when many previously forbidden literary works were published for the first time, such as Solzhenitsyn's *One Day in the Life of Ivan Denisovich* or *Cancer Ward.* Khrushchev made the publication of these novels possible.[7] I remember reading them for the first time, with my eyes wide open. It was a period of hope: it seemed to us that things started to change, that our society was thawing out. Then Khrushchev gave his famous speech at the Twentieth Party Congress, in which he denounced Stalin. At the Mendeleev Institute, they invited Komsomol and party members to a closed meeting, and read it to us. It was a long speech, and they read it for 3–4 hours. I was shocked when at the end someone yelled from the back of the room: "Is this a provocation?" Naturally, students started asking their teachers questions. And they had a lot of questions. They wanted to know why their teachers had remained silent and whether they had known about all these things. I remember we had a class at the Mendeleev Institute, called "The Foundations of Marxism-Leninism." Our teacher was a rather dumb, poorly educated person. One day he called on Valery Legasov and asked him if he had read the yesterday's issue of *Pravda*. Valery responded: "I don't read *Pravda*." I could hear other students gasp. "How can you not read *Pravda*?" the teacher asked. "Do you read any newspapers at all?" And Legasov said: "I don't read *Pravda*, because it is all lies." We thought that Valery would be immediately expelled from the Komsomol, maybe even from the institute. But nothing happened.

Did you start working on the "big nuclear bomb" at the Mendeleev Institute?

Not exactly. We worked on the chemistry of the nuclear bomb. You had to extract and split uranium, and I was conducting the experiments. At the time, few people recognized how dangerous radiation was. As a result, I got exposed to a substantial dose of radiation, but thank goodness it didn't lead to any serious health problems.

Oleg Nikolaevich, can you tell me about your friendship with Kronid Liubarskii?

Well, I was one of the few people who was friends with Kronid Liubarskii in Chernogolovka. I also introduced him to Rustem Liubovskii.[8] Kronid Arkadievich was a brilliant translator, and he asked me to help him translate a book called *Chemical Evolution*. The book was written for a wider audience, and some paragraphs were quite complicated to translate. I told Kronid that my English was not good enough, but that I would introduce him to my friend, Rustem Liubovskii. Rustem was also struggling with the translation, but they started working on it anyways. The translation came out in 1971 with both of their names on it. Rustem and his wife, Rimma, and my former wife and I used to spend many evenings at Liubarskii's apartment. I was most interested in the numerous books that he had. He allowed me to take them home. Sometimes I gave these books to a few other people to read.

Are you talking about samizdat books?

Well, yes, the so-called samizdat. One day he asked me to take some of these books and to hide them in my apartment. I guess he felt some danger looming. He brought the entire suitcase full of books to my place, and I put it under my bed. Shortly before his arrest, Kronid took these books back, even though I protested, saying that they were safe with me. Soon after that, they [the KGB] searched his apartment. Kronid Arkadievich and his wife, Galina Il'inichna, were not allowed to leave, but their daughter, Vika, could go. So she came running to my place and said: "Uncle Oleg, they are searching our apartment." The next day, Kronid Arkadievich came to see me. He brought Galina Il'inichna with him. They knocked on my door and when I opened it,

Kronid said: "Are horrible people like us still welcome in this apartment?" I said: "Well, of course, Kronia, how can you say something like this?" We started talking, discussing the possibility of him leaving town. But he said: "Where would I go?" They arrested him shortly after that. And several months later, I received a phone call, around 4 a.m. in the morning. The phone call was from the KGB and they were calling to summon me for an interrogation. My first interrogation took place in Noginsk.[9]

Did they explain why you were summoned?

Yes, they had questions about Kronid Arkadievich. They wanted to know whether I had visited Kronid's apartment and what books I had borrowed from him. I told them that we mostly discussed science. It was a senseless conversation. At some point, I remember they were telling me that we were part of the rotten intelligentsia, who didn't appreciate that the KGB was simply fulfilling its duty.

How did they find out about your friendship with Liubarskii?

Someone must have told them.

Did many people at the Branch of the Institute of Chemical Physics know Liubarskii?

No. Although once I invited Kronid to give a lecture about Mars at our institute's seminar. He studied that topic closely and he had a wonderful collection of photographs. There were about sixty people at the seminar, maybe one of them mentioned it to the KGB. I am not sure. Kronid didn't work at our institute. He taught physics and astronomy at a local school.

How did you meet Kronid in the first place?

I first met Kronid Arkadievich at the Moscow Planetarium. He came from Simferopol, where they had a society of young astronomers, and he was getting ready to take his entrance exams at Moscow University. Kronid Arkadievich was several years older than me, although he appeared to be much more mature. He was extremely well educated, and immediately drew attention to himself. As far as Galina Il'inichna goes, I met her at the Planetarium, too. We met several years earlier and became close friends.

Did Kronid Arkadeivich and Galina Il'inichna meet at the Planetarium as well?

No, they met at the Mechanical and Mathematical Department of Moscow University, where they were both students. After graduating from Moscow University, Kronid accepted a job at a small observatory in the mountains in Turkmenia. Galina Il'inichna joined him several years later. We lost touch for a while. They returned to Moscow in the mid-1960s, after Kronid got into graduate school at Moscow University, where he did his doctoral work on meteors. They didn't have their own place and had to live with Galina Il'inichna's family for a while. Her father worked at the KGB, by the way. Her father was a good man, a patriot of his country. He fought in the war, participated in various landing operations behind the frontlines. It was a miracle he survived. The Liubarskiis moved to Chernogolovka sometime in the late 1960s. As I said, we had lost touch for a while, but they found me shortly after they moved here.

Did you know that when Kronid was a student at Moscow University, he was one of the organizers of the open letter in support of Vladimir Pomerantsev's article "On Sincerity in Literature."[10]

I might have heard about it from Galina Il'inichna, since at the time I was closer friends with her than with Kronid Arkadievich. My friendship with Kronid was different: he was an authority figure for me. I was especially interested in the samizdat books he had. But that was later, after they moved to Chernogolovka. They received their own apartment here, and my ex-wife and I visited them frequently, almost every night. Kronid had a ton of books, and I borrowed and read a lot of them. The central among them was *The Chronicle of Current Events*. It was published on extremely thin paper, and Kronid participated in its production. *The Chronicle* gathered and published information about political trials and political prisoners in the Soviet Union. It also offered support to the families of political prisoners. For the KGB, it was the most criminal and dangerous publication, and they did all they could to destroy it. Kronid Arkadievich also had many good literary works at home. I read many of them and we discussed them afterwards.

Did you borrow The Chronicle *as well?*

Of course, I did. I had no idea there were so many political prisoners in our country. Sometimes Kronid's dissident friends from Moscow came to visit him in Chernogolovka, but he didn't have enough room to accommodate all of them in his apartment. So some of them stayed at my place. Grigorii Pod'iapol'skii, for example.

Could they enter Chernogolovka without any restrictions?

Yes, they could come and go as they pleased. There were no special restrictions.

There was a funny story, by the way. Kronid was an expert on Mars, and he worked as a consultant on a Soviet documentary about Mars. The film came out after Kronid had been arrested. So he was already in jail, but his name still appeared in the film credits. I guess they forgot to cut it out.

Do you know why Liubarskii became a dissident in the first place? After all, he could have joined the ranks of the scientific elite, like many others.

You know, I once had an open, honest conversation with him about this. He told me that he became less interested in academic research and was drawn to journalism instead. He translated many scientific works, like *Chemical Evolution*. But his interests shifted away from science. Around the same time, he became acquainted with many famous dissidents in Moscow, and became friends with some of them. These relationships must have had a significant impact on Kronid's views. He wanted to write about that. This is how he got involved in publishing samizdat: samizdat provided information that was not accessible anywhere else in the Soviet Union. Kronid was never a leader of the dissident movement. He didn't participate in any open protests or demonstrations. But he shared the dissidents' views and he fought for them. I neither understood Kronid nor shared his views. My head was full of science back then. I used to tell him: "You and your friends criticize everything, and a lot of your criticism is fair. But you should have some positive suggestions, too." One time Kronid got really upset with me and accused me of being too materially minded (laughs).

Did the dissidents have a positive agenda?

It seemed to me that they didn't, at least back then. They demanded that the Soviet authorities respect the Constitution of the USSR. They were convinced that if the Soviet regime had respected the Constitution, things would have been fine. But I thought that it was insufficient. Yes, we had a good Constitution, but we needed some constructive ideas, too. The dissidents wanted to destroy the existing order, or something like that.

Would you say that they were kinds of revolutionaries?

Maybe they were.

And what were your political views at the time?

I found all this fascinating. My friendship with Kronid allowed me to see the world around me in a completely different light. I told you that I had joined the Komsomol under pressure. Later they kept asking me to become a party member, too. But I found various ways to decline. Ironically, it was shortly after Kronid's arrest in 1972 that they asked me to join the party for the last time. Everyone in town knew that we were friends. But the head of our institute's party organization came to me anyways and said that now I would certainly have to join the party. He was a naive person. I laughed in his face. I told him that my real passion in life was alpinism and that I would be ready to join the party only once I had climbed the highest mountain in the Soviet Union, called "Communism Peak" (laughs).

How many times did they ask you to join the party after you moved to Chernogolovka?

They asked me at least three times.

And why did you refuse?

I was convinced that the Communist Party was a bizarre organization. It was like a compote, in which some pieces of fruit were edible, but others were not. I respected some Communists, because they were good people. But I thought that other party members were scoundrels. Yet they were all members of the same party, required to vote in the same way. I didn't want to support the decisions and resolutions I disagreed with. They called it "democratic centralism." As a party member, I also would be expected to denounce other party members if I suspected them of any anti-Soviet activity. Why would I want to do that?

Things became more complicated when I started thinking about establishing my own laboratory. It was in the 1980s. Fortunately, it was already during the Gorbachev era, and they softened this requirement for lab directors.[11] Still they summoned me to the meeting of the institute's party bureau. "How could they trust me with directing a lab, if I wasn't a party member?" they asked. Still most people voted for me, and the party bureau had no choice but to approve my candidacy.

Does it mean that you never joined the Communist Party?

I never did. Despite this, I have directed my own lab for twenty-six years now. I was also head of a large scientific department, which employed almost 200 people, for sixteen years. I resigned as a head of the department three years ago, but I still have my own lab. Even though I was not a member of the Communist Party, they asked me to participate in various committees of party control. Some of our work was actually useful. I met some wonderful, honest people there. I didn't want to join the ranks of the party, because I didn't believe I could change anything. I was convinced that party membership could make me corrupt.

And what about Kronid Liubarskii's political views? How did they differ from yours?

Kronid didn't oppose socialism. Like most dissidents, he was a human rights defender. He wanted the Soviet state to respect the civil rights of its citizens, such as the right to vote and to publish freely, and the right to participate in demonstrations.

These rights are part of the Western tradition.

These rights belonged to all humanity, but were openly violated in the Soviet Union. This is why the dissident movement originally emerged. The dissidents' human rights activism inevitably became a political matter in the USSR. Their protest against the Soviet invasion of Czechoslovakia, for example, was largely a political protest.

And Kronid focused mostly on human rights, correct?

It was more complicated than that. I don't think, for example, that he was in favor of a one-party system. We didn't discuss this directly, but it was implied. Once I read a wonderful book by Arthur Koestler, *Darkness at Noon*. I borrowed it from Kronid's library. Koestler was a brilliant writer, and one time I got bold and gave a talk on this book at the philosophical seminar at our institute. I didn't openly criticize socialism or the Soviet government, but since I sympathized with Koestler and Orwell, it was obvious that I disapproved of Big Brother and all that.

How old were you when you started reading books like this?

I started reading them after Kronid moved to Chernogolovka. I usually could only borrow these books for several nights, but I was completely captivated by them. These books drastically changed my worldview. Yet Kronid never tried to convert me, and I didn't start seeing the world from his perspective. In a sense, Kronid reminded me of Kaminskii, whom I mentioned earlier. He didn't want me to start thinking like him. He gave me these books so that I would have an opportunity to read them and start thinking on my own. He believed in the power of self-education and free will. Reading and critical thinking were key to the formation of self-reliant individuals, who could not be easily influenced or broken by the KGB. Ironically, one of the books I borrowed from Kronid was about how to behave during interrogations (laughs).

Did Kronid give it to you?

No, he just let me look through his library, so I found this book on one of his shelves. I asked his permission to borrow it, and Kronid said: "Oleg, it's too early for you to read this book." (laughs). It was shortly before his arrest. After the arrest, I was summoned for several interrogations. The first one took place at the KGB office in Noginsk and it lasted for five hours. Since I had read that book, I knew that during interrogations they write reports, which they ask you to sign at the end. So as soon as I walked in, I asked them if they were going to write a report. They asked me how I knew about that, hinting that this proved that I had read that book. I said I had not read any books; I simply wanted to know what my rights were. They told me it would be a simple conversation, no reports. My next interrogation took place in Lefortovo Prison in Moscow. This time it went on for twelve hours. I remember they laid out different books on the desk. They wanted me to tell them which books I had borrowed from Kronid.

When did these interrogations take place?

Well, Kronid was arrested in January 1972. They summoned me for interrogations several months later, so they took place sometime in March. What rattled me was that they called me in the middle of the night, at 4 a.m. After that I never really liked early morning calls.

Were you afraid?

No, I was not. They called me, so I had to go. I was a bit shocked, though, since they called me so early in the morning, when I least expected that. Now that you ask me this question, I believe that one of my interrogations took place around March 8, which is International Women's Day. I remember this, because when I returned to the institute, there was a celebration in our laboratory. But Kronid Arkadievich didn't try to convert me to his beliefs.

Did conversations with him open your eyes to things you had not noticed before?

Not really. What opened my eyes was reading the books, which I had borrowed from him. I already told you that sometimes he would ask me to host some of his Moscow friends in my apartment.

Did you know they were also dissidents?

Of course, I did.

Were you afraid that this might get you in trouble?

I never thought that it might lead to an arrest. I also had a security clearance, which I could have lost. But back then we separated these things. Even Sakharov, who was a prominent dissident scientist, never publicly revealed any classified information. I was a scientist, too, and I could relate to that. Moreover, I led a double life, in a sense. On the one hand, I had my work and I read all these books. On the other, my true passion was alpinism, and I lived in that world primarily. It was a world full of passion and incredible adventures. Ascending these mountains was akin to a religious experience. It made me realize the imperfections of our world. I loved my job, but I lived for climbing the mountains. I enjoyed talking to the dissidents, of course. They were wonderful people. I remember when Kronid was finally released from prison. They let him visit Moscow for one day. There was a meeting at someone's apartment in Moscow, and I rushed there to see him, even though we knew that the KGB was watching our every step. Later, I also visited Kronid and Galia in Tarusa, during their short exile there. We spent the whole night talking, drinking tea, and laughing, even though we knew that they were recording every word we said. Kronid had a big poster in his bathroom, which said "Thank you for listening!" When I returned to Moscow, a police officer stopped me at the railway station. He asked me to show him my passport. He wrote down its number and let me go. When I went inside the metro, they stopped me again, but then let me go.

Could this incident have had any negative consequences for your career?

Yes, of course, it could. Fortunately, it didn't affect my research in any serious way. I always enjoyed professional freedom. They trusted me. Yet when I applied to travel abroad to a scientific conference, they turned me down.

Let us come back to this in a little bit and finish up with Liubarskii first. Did you and Kronid Arkadievich do any joint scientific research?

We were both interested in studying Mars. I told you that I published my first scientific paper when I was a member of the Moscow Planetarium. What I didn't mention is that it was about dust storms on Mars. All my life, I collected all kinds of publications about Mars. And Kronid was an expert on the planet. He even served as a consultant on a film about Mars. We discussed recent findings and publications on Mars, but no, we didn't have any joint publications. Besides, Kronid was aware that some of my research was classified, and he never asked me about it. Sometimes he joked, saying that my work helped build up the military might of the Soviet state. I joked back. You see, the Soviet Constitution of 1936, or the Stalin Constitution as they called it, was allegedly the most democratic constitution in the world. If the

Soviet regime had only respected the Constitution, if it had respected freedom of speech and freedom of the press, maybe then people would have started thinking differently earlier and they would not have tolerated a one-party state. The dissidents demanded that the Soviet state respected the Constitution. There were not many truly heroic figures among them. Andrei Sakharov was, of course, at the forefront of the dissident movement. But he was a unique person: a Nobel Prize winner. Later, they named a street after him.

Do you think that overall Soviet society benefited from the dissident movement or not?

Soviet society clearly benefited from dissident activity. Dissidents made people think. It is one thing when your education is limited to reading *The Short Course of the CPSU*.[12] It is completely different if you have access to all this unpublished literature. The dissidents made this possible.

Did the dissidents aspire to offer an alternative to the official ideology?

Not exactly. The official ideology stated that there was a one-party state in the USSR and that the party was above politics and elections. The dissidents didn't really want to destroy the Soviet regime. They were more like the nineteenth-century Decembrists.[13] The Decembrists didn't really want to abolish the tsarist regime. They simply wanted to have more freedom in Russia, they wanted Russia to be more like France after the French Revolution.

And what do you think inspired Soviet dissidents? Was it Khrushchev's Thaw?

Perhaps. These couple of years were extremely important. It was largely thanks to Nikita Sergeevich Khrushchev that Solzhenitsyn's works were first published in the Soviet Union. Khrushchev's speech at the Twentieth Party Congress was even more significant. It had a tremendous impact on our formation. I remember I had only one thought, as I was leaving the meeting where they read the speech: "I lost my ideology." I felt empty. Everything that I had believed in growing up disappeared without a trace. I think many people felt the same way. There was nothing left to believe. There was no one to trust. It was obvious to us that this ideology was corrupt. Eventually, I abandoned it. I came to God and got baptized.

When were you baptized?

It happened many years later. My mom died in 1994, and before she died she asked me to get baptized, so I did. I read a lot of different books about religion. I accepted a completely different worldview. It is important for a human being to have faith. It gives them structure in life.

But going back to Kronid, one of my interrogations took place in Lefortovo. It lasted the entire day, and the investigators did all they could to make me talk. At some point, they told me that Kronid Arkadievich was imprisoned in the same building, only several floors below me. They said that if I didn't start talking, they would bring him in and make us confront each other. To this I responded that I would love to see Kronid Arkadievich and that it would be wonderful if they brought him in. They didn't do it, of course. Instead, they sent in another investigator, who said to me that I should tell them everything simply to get it off my chest and to ease my soul, so to speak. I remember I looked at him and said: "It seems there is a big difference between your organization and Church." He didn't like my response, but they let me go after that.

Was that your last interrogation?

No, they summoned me to Lubianka as well, although it was already late spring or early summer, and I was getting ready to go mountain climbing. And then there was Liubarskii's trial in the fall. But I was not invited.

Were you not invited or not allowed to attend the trial?

Both. I went there anyways, of course, but there was a huge barn lock on the door of the courthouse. There was also some silly sign that said the building was under renovation.

Did you go by yourself?

No, I went with Rustem Liubovskii. The trial took place in an old courthouse in Noginsk. They brought a group of scientists from Chernogolovka to attend the trial, but Rustem and I were not part of that group. One of the scientists who attended the trial was the secretary of our institute's party organization. Later, he invited me for a walk to discuss what had happened inside the courthouse. I think he received permission from above to talk to me about it. Liubarskii was accused of distributing samizdat literature, which violated Article 70 of the Soviet Penal Code. Kronid admitted that he had given books to other people. This is why they summoned me for interrogations in the first place: they needed to prove that Liubarskii had given me samizdat books to read. I denied it, of course, but eventually they found other people who admitted that he had given them books to read. This is why they let me go at the end of the day.

Could they have accused you of distributing samizdat as well? You mentioned earlier that you had given these books to a few other people.

They could have, if they wanted to. But their main goal was to stop the production of *The Chronicle of Current Events*, which they had found during the search in Kronid's apartment.

Does it mean that at that point the KGB was only interested in open dissidents? Were they no longer looking to arrest people who privately criticized Soviet power?

Correct—it was not the 1930s any more. They no longer broke people's jaws or crushed their teeth during interrogations. It was more civilized. The KGB had a goal to destroy samizdat, and they were chasing after people, like Kronid, who helped produce samizdat. I don't know how successful they were because people still read samizdat, sharing it privately with their friends.

But the KGB did manage to break some dissidents, like Petr Iakir. They knew how to break people; they were well trained to do this.

Anyways, Rustem and I went to the courthouse in Noginsk to see Kronid's trial. He was our friend and we wanted to be there. It was cold outside. Since they didn't allow us to enter the courthouse, we had to walk around the building. We ran into Andrei Dmitrievich Sakharov and his wife Elena Georgievna Bonner, who also came to support Liubarskii. We stayed for several hours.

Did scientists who were present at the trial discuss it at work later?

Absolutely not.

It seems a bit strange that they didn't talk about it. After all, Chernogolovka was a small town, and people must have known about the arrest and the trial of Liubarskii.

Nothing like this happened. I think this was a deliberate effort on the part of the local administration. They wanted to present Liubarskii as an outsider, who ended up living in Chernogolovka largely by accident. Indeed Kronid didn't interact with many local scientists. His main relationships were within Moscow dissident circles. It was important for the local administration to separate the Liubarskii Affair from the local scientific community. They didn't want the KGB to think that Chernogolovka was a hotbed of dissidents. After all, it was a closed scientific town. Thus it was more convenient for them to say that Liubarskii was an outsider, who had taught astronomy at a local school, and that is how he got his apartment in Chernogolovka. Besides, it was a closed trial. Those who attended the trial had to vouch that they would not tell anyone about it.

Moreover, people still remembered what had happened in 1968, after the Soviet invasion of Czechoslovakia. Several people at our institute discussed those events during their break at work. Someone must have overheard them and reported on them. Because several days later, one of them was called into the institute's party organization office. We simply were not allowed to discuss such things.

How did you find out about the events in Czechoslovakia? Did you read Soviet newspapers?

We listened to Radio Liberty. Everyone would turn on their radio sets at night and listen.

Not everyone. Some of your colleagues told me that they didn't listen to foreign broadcasts at all.

Why not? Of course, it was challenging to find them, because the KGB jammed the broadcasts. It was not as bad here in Chernogolovka as it was in Moscow. The quality of the broadcasts was poor, but at least we could hear them.

Do you have a sense of whether the Soviet invasion of Czechoslovakia in 1968 made your friends and colleagues more critical of the Soviet regime?

I don't think it did. Soviet propaganda worked hard to brainwash people, claiming that the reforms in Czechoslovakia were the product of American and West German intelligence operations. Listening to Radio Liberty undermined what Soviet propaganda was saying. But even then it was hard to know which sources were more reliable. I listened to both Soviet and foreign broadcasts, and I tried to make sense of them. Besides, when this happened, in August 1968, I was ascending the Pamir Mountains. There were several expeditions not far from ours. One of them was from Czechoslovakia. The Czechs were trying to find Czech radio stations. They were somber. I remember we made a big fire and sat around it, singing songs. They sang in Czech, and we sang in Russian. And we had some bizarre conversations about what would happen next. The Czechs were afraid that they could be detained and possibly exiled to Siberia. There were some scientists among us, who came from a closed nuclear town in Siberia, and they tried to console the Czechs, saying that Siberia is not that bad. So you see, we had different sources of information, apart from samizdat. The problem was how to process and reconcile these different pieces information. It was not at all clear. At the moment, we could not fully appreciate the significance of what was happening in Czechoslovakia. The same was true for the Hungarian uprising in 1956. I once went mountain climbing with a scientist who visited Hungary during the uprising. He was telling us that the Soviet invasion in Hungary was justified, because they had to stop the killings of Communists there. He almost convinced us, but then there was Radio Liberty.

It is somewhat ironic, but sometimes I think that under Brezhnev scientists trusted the government much more than they trust the present-day Russian government. It is important to understand one thing: people came to Chernogolovka and, sooner or later, they received separate housing here. Take me, for example. I moved five times around Chernogolovka, since I got here. At some point, I lived in a four-room apartment. So first, the Soviet government provided scientists with housing, which was extremely important. Second, we had good infrastructure here, including kindergartens and schools. Third, even though we had small salaries, we made enough money to enjoy decent living. There were not many things to buy in stores anyways. Moreover, Fedor Dubovitskii secured privileged food supplies for us, especially in the beginning. He made sure that we had various foodstuffs here, including red caviar.

I heard about red caviar from some of your colleagues, too. Surprisingly, they barely mentioned any other foodstuffs.

Well, in 1962–1963 there was a small grocery store located inside the Polymer building of our institute. I had just recently moved to Chernogolovka, and I passed by the store every morning, on my way to work, and stopped to see what was available. It was mostly basic foodstuffs such as bread, milk, and meat. There were still few people living here, so we could buy food without too much trouble. There was also a system of rationed food parcels (*systema zakazov*), which allowed us to buy something special for the holidays.

You see, even the Soviet state took better care of its scientists. In fact, the Soviet state did a much better job than the Russian state does today. True, we didn't have the best scientific equipment or reagents for our experiments, but at least the state allocated money for buying them. The post-Soviet state stopped investing in science. It pays us meager salaries. If we need money to conduct research, we have to apply for grants. There is not much you can do without grants. As a result, Soviet scientists had much higher trust in the government than Russian scientists have today. You should also keep in mind that people who worked at the party organizations or trade unions—these were all our people. They were normal, decent people, who helped us in our everyday lives and work. So even though some scientists listened to Radio Liberty and read samizdat, it could not have had any serious impact on their thinking.

Chernogolovka was also a closed town, and some research conducted here was classified. This created a special atmosphere in the town. Once a KGB representative came here to give a lecture about how important it was to be politically educated and to avoid reading literature that slandered the Soviet regime. He wanted us to participate and ask questions. But the audience remained silent. We didn't discuss this among ourselves afterward either.

What year was that?

Around 1965.

Why didn't you discuss it among yourselves?

I didn't pay too much attention to what he was saying. We were also afraid that someone might overhear us and report on us. My inner voice always told me that such conversations could get me in trouble. I didn't know my coworkers that well, so I could not be sure that they would not report on me. Besides, there was a veil of secrecy surrounding the work of scientists who lived in closed cities. We didn't even know about Sakharov's contributions to building the nuclear bomb. We barely heard about Arzamas-16, or present-day Sarov. A lot of the information remained classified. Many people from my cohort at the Mendeleev Institute were employed in classified scientific cities. I was an exception, in this respect, since I went to work in Chernogolovka.

How did you end up in Chernogolovka, by the way?

It was almost by accident. I was finishing up my studies at the Mendeleev Institute, and I was looking for a job that would also provide me with housing. The housing situation in Moscow was terrible. I knew some people who had been working in Chernogolovka already, and they invited me to come visit them. I did and I really liked it here. Later they introduced me to

Aleksandr Evgenievich Shilov, who was recruiting scientists for his department. So I became one of his first employees.

Oleg Nikolaevich, some of your colleagues told me that there was more intellectual freedom in Chernogolovka and that they could breathe more freely here. Do you agree with that?

Not really. There was more freedom in Akademgorodok in Siberia, or even at Moscow research institutes. Here we invited Vladimir Vysotsky to give concerts a couple times.[14] But even then they had to cancel his second concert. So I would not say that we enjoyed more intellectual freedom. We simply lived and worked here. We were busy conducting research, which we found captivating. After all, back then science was on the rise in the Soviet Union.

Would you not call Chernogolovka "an oasis of intellectual freedom" then?

No, not at all. What kind of intellectual freedom are you talking about? We constantly had to attend mandatory political seminars, during which we had to blindly approve various party resolutions. Sure, we could discuss how to best advance science. But all political discussions were off the table. The goal of these political seminars was to ensure that all employees thought alike.

And did it work?

It most certainly did.

Was there any diversity of thought in Soviet society after Stalin's death?

Well, many years had to pass for this to happen. You and I discussed the Thaw a bit. It was a brief period, when people finally got access to various books, which suggested that things were not as good as Soviet propaganda had claimed. We found out about the repressions.

It looks like, on the one hand, Soviet scientists read samizdat and listened to foreign broadcasts. Yet, on the other hand, they remained loyal to the Soviet state. How do you explain this seeming contradiction?

You see, the scientific intelligentsia was not yet mature. When you read Solzhenitsyn's *First Circle*, you find out that many intelligent people continued to support the Soviet state even after they had been imprisoned. Scientists in *sharashkas* continued to design wonderful air planes, because they wanted to contribute to the well-being of their Motherland.

When I was younger, I thought that I was extremely lucky to have been born in the Soviet Union. Vladimir Mayakovsky was my favorite poet. Nowadays I only read his early poems, but I used to read and trust everything he wrote in support of Soviet power. You should not forget that there was a lot of rhetoric aimed at raising patriotic feelings, too, especially surrounding the war. Overall, the Soviet intelligentsia remained loyal.

Did their loyalty come to an end at some point?

I can only speak for myself. For me, the official ideology collapsed after Khrushchev's speech at the Twentieth Party Congress. I realized that they had lied to me all these years and that I could not trust these political leaders any more. It was obvious that Khrushchev had been involved in the repressions, but he stayed silent, because he was afraid. What kind of a Communist are you, if you are afraid? Under Brezhnev, we came to recognize that the economy was stagnating and that it would lead to an economic disaster at some point. Some of the people teaching political seminars actually talked to us quite openly about the economic problems in the country. But no one thought about dismantling the Soviet state. Maybe repairing it. So when Gorbachev came to power, we believed—at least I believed—that, finally, here is a political leader who had a new way of thinking. He told us that we had to be patient, and things would work out. He said it would take several years. Many years have passed since then, but things have only gotten worse. I don't know much about it, but I am struggling to understand why China, for example, was able to succeed economically, approaching America today, but we failed. Why did we choose such a strange path?

Did Soviet scientists become more skeptical or even disillusioned with the Soviet state?

I am not sure this is the right question to ask. Even if they did change their views, this transformation had little to do with politics. You see, whenever I thought about problems plaguing the Soviet state, I concluded that the economy was not working. There was a time in Soviet history when a totalitarian economy was a necessity. Stalin was the greatest villain and scoundrel, but he got one thing right: cadres decide everything. We could buy the modern machinery from the Americans, but at the end of the day cadres decided everything. The Soviet state invested a lot into these much-needed human resources. It invested a lot in creating the scientific intelligentsia. And we were proud to be part of it. It was prestigious to be a scientist employed at one of the research institutes of the Academy of Sciences, to serve as a head of laboratory or a department. Yes, I might have been critical of certain things, but this was secondary. We had common goals with the Soviet state. It was obvious at the time that *our* interests overlapped with the interests of the Soviet state. Today this is not the case. So if you are looking for a transformation of beliefs, it only happened within the last 10–15 years, when the interests of scientists and the state grew apart.

Are you saying that scientists and the state had shared goals until the very end of the Soviet Union?

Yes, they did. They had shared interests during the Brezhnev era. Despite some concerns with the course of Soviet development, scientists had no major disagreements with the Soviet regime. We knew that we lagged behind in certain spheres, but we thought we could simply repair this. We never thought about restructuring the entire state and society. True, there was no freedom of speech or freedom of discussion, which made it difficult for new ideas to emerge. But we didn't associate this with the existence of a one-party system. It was only under Gorbachev and Yeltsin that we first started thinking about it.

Does it mean that overall scientists didn't suffer from the lack of freedom?

You are right about this: they didn't. All these democratic ideals. . . . Russia was never a democratic country. It is a very traditional country, and it takes time to change this. If you remember, in the Bible, Moses and the Israelites spent forty years wandering in the desert until the last person born in slavery died. The same is true for us: we will keep looking for a path forward until the last person born under Communism dies.

Yet, by the late 1970s few people seriously believed that it was possible to build Communism in the USSR.

Well, Khrushchev was the last fool who promised that the next generation would live under Communism. He said that we would catch up with and surpass America in the production of meat and milk. We didn't believe him, of course. He was not a well-educated man. Despite this, he got many things right. Still, I am convinced that the Communist Party had a strong grip on power until the Soviet collapse. People remained largely brainwashed. But you should not throw the baby out with the bath water. Many Communist ideals were really the ideals of all of humanity. We should not lose sight of that.

Do you think that economic problems were one of the main reasons for the Soviet collapse?

Yes, I do. A totalitarian economy worked only up to a certain point. It took Americans 200 years to industrialize. But we had to carry out the industrialization much quicker. There were only seventy years of Soviet power. Even though we managed to catch up in our industrial development, we lagged behind technologically.

Oleg Nikolaevich, let us change the subject. Did you travel abroad, especially to the West, in the Soviet era?

Yes, I did.

When was the first time you went abroad?

Well, it was a requirement that before one could travel to a capitalist country one had to go to a socialist country, like Poland or Bulgaria. My first trip was to Bulgaria as a scientific tourist. During that trip, I noticed that they lived better than us and that they had a higher quality of living. I also saw that they no longer liked the Soviets. I traveled to the GDR, too. They were much more technologically advanced than we were.

Did you go abroad before or after the Liubarskii Affair?

Before Kronid, I went to Bulgaria once. Then I applied to go to Poland, but they denied me permission to leave the country. At the time, there was a long procedure to receive an "exit" visa. I had to pass an interview at the October regional committee of the CPSU. They asked me some ridiculous questions.

One of them was whether I knew about some local initiatives at a factory in Poland, which I had never heard of. I said I didn't. So they started lecturing me, telling me that I should go read about these initiatives first and then come back for another interview. The whole procedure was revolting. What gave these old people, who were old enough to have met Lenin, the right to tell me off? I was going to Poland to do scientific work, not to propagate Soviet power. After they started lecturing me, I got up and left, slamming the door behind me. Later, they tried to convince me to come back for another interview, but I refused.

Was the purpose of the interview to prepare you for your trip ideologically?

I was prepared. If you are talking about my political views, I always knew that it was better to leave my political views at home. In private, I could argue with Kronid Arkadievich about the Soviet regime and the Soviet Constitution. But when I came to work, my attention was completely focused on science. I was an ambitious young man who aspired to revolutionize my field of research. And such commitment required time and dedication. At the end of the day, I managed to accomplish quite a few things. My life was not all a waste.

Did scientific research present you with an opportunity for self-fulfillment?

Yes, absolutely. And when my children grew up, I told them that they should emigrate to the West, if they wanted to have opportunities for professional self-fulfillment.

Did you think about emigrating from the Soviet Union?

No, I thought it was too late for me. When we were young, we were not allowed to leave the country. Later we came to realize that to be successful professionally, we had to emigrate. Most people who left were not driven by material considerations. They were ambitious people, looking for opportunities for self-fulfillment.

It is curious: on the one hand, you are saying that Soviet scientists remained loyal to the Soviet regime until the end and that the official ideology still had a strong grip on them. On the other hand, just now you said that you were encouraging your children to emigrate, because they could not fulfill themselves professionally in the USSR. If scientists continued to believe in the Soviet project, why did they encourage their children to emigrate?

Well, this is not exactly right. My generation . . . you see our ideology broke down after Khrushchev's "secret speech." After that, it became obvious that we had to look for answers and self-fulfillment outside of the state and party structures. People started thinking in different ways. This individualism gradually began to replace collectivism. Before we felt that we were all part of the collective. But once this broke down, we became more and more individualistic. I told you about Kaminskii, the polar explorer. He taught me to be self-reliant. Don't wait for anyone to help you. If they do help, thank

God. If they don't, you should work hard to accomplish your goals without anyone's help.

Would you agree then that by the time Gorbachev came to power, there was much more diversity of thought in Soviet society than, say, under Khrushchev?

It is a complicated question. What do you mean by diversity of thought? I told you that the official ideology broke down under Khrushchev. You could not keep believing in Communism after what you heard. So we started looking for truth elsewhere. If you open these bookshelves, you will see that they are full of books on philosophy and political philosophy. I read them all. I had to pass a philosophy exam when I was doing my graduate work. The list of books they gave us was very restrictive. I was bored to death reading *The Short Course of the CPSU*. Of course, I read other books instead. I read Sigmund Freud, for example. Solzhenitsyn's *One Day in the Life of Ivan Denisovich* was published under Khrushchev. It was incredible! Then there was *Novyi Mir*, a literary magazine that provided an outlet for the intelligentsia. I was a subscriber and used to collect all the issues of this magazine. If you read it, you will understand the philosophy and the world-view of my generation. Then there was *Literaturnaya Gazeta*. There were public poets' readings at the Polytechnic Museum in Moscow. We were the regulars at all these events. Yevtushenko, Voznesensky, Akhmadulina. . . . All the new poetry and prose that appeared in *Novyi Mir*—they shaped our thinking. And all the samizdat books I read at Kronid's place. All this played an extremely important role in my formation and the transformation of my beliefs.

NOTES

1. The Moscow Planetarium was founded in 1929. It had modern astronomical equipment and hosted study groups and lecture series by prominent Soviet scientists, among other things.

2. The Communist Youth League (*Komsomol*) was a political youth organization in the Soviet Union, established in 1918. It was the final stage of young Communists' education; Komsomol membership was an essential requirement in applying to a higher education institution in the USSR.

3. Lavrentii Beria was chief of the Soviet secret police under Stalin. After Stalin's death in March 1953, he became the most powerful man in the Soviet Union, heading the Soviet government and the Ministry of Internal Affairs. Beria was removed from power in a coup organized by Nikita Khrushchev and supported by the military. He was arrested in June 1953, put on trial, and executed in December 1953.

4. The Doctors' Plot was a part of the anti-Semitic campaign during late Stalinism. It was devised by Stalin during the last months of his life and was directed mostly

against Jewish doctors, who were accused of trying to murder high Soviet government officials.

5. It was highly prestigious to be elected a corresponding or a full member of the Soviet Academy of Sciences.

6. Igor Evgenievich Tamm was a prominent Soviet physicist and a winner of the 1958 Nobel Prize in Physics. He was one of the theoretical physicists who contributed to the construction of the Soviet thermonuclear bomb.

7. While *One Day in the Life of Ivan Denisovich* was first published in *Novyi Mir* in 1962, Solzhenitsyn's *Cancer Ward* came out in 1966, two years after Khrushchev's removal. It was first distributed in samizdat.

8. In his interview, Rustem Liubovskii recalls that it was Isai Braginskii who first introduced him to Kronid Liubarskii (see Chapter 5).

9. Noginsk is a city thirty-six miles to the east of Moscow and thirteen miles to the south of Chernogolovka.

10. The essay "On Sincerity in Literature" was published in *Novyi Mir* in December 1953. It criticized the lack of sincerity and the varnishing of reality in socialist realism. The essay marked the beginning of the Thaw in Soviet literature.

11. Efimov is confusing dates here: his laboratory was established in 1983, while Gorbachev came to power in 1985.

12. *The Short Course on the History of the CPSU (b)* was published in 1938. It was commissioned by Joseph Stalin and contained Stalin's own exposition of the fundamentals of Marxism-Leninism. It became the canonical text of Stalinism and was required reading for every schoolchild or student in the Soviet Union.

13. The Decembrist movement was a revolutionary movement in Imperial Russia. It was composed of army officers and members of the Russian nobility. A lot of them participated in the Napoleonic Wars and were exposed to the tradition of Western liberalism during their military campaign in Western Europe. Upon their return to autocratic Russia, they staged an uprising in December 1825, refusing to swear their allegiance to the new tsar, Nicholas I, and demanding a constitutional monarchy. The uprising was quickly suppressed by Nicholas I. The Decembrists were arrested. Some were executed; others were imprisoned or exiled to Siberia.

14. Vladimir Vysotsky was a Soviet actor and a semiofficial singer poet, who indirectly criticized the Soviet political and social system in his songs. He was extremely influential and popular among the Soviet intelligentsia in the 1970s.

Selected Bibliography

Abrams, Lynn. *Oral History Theory*. London; New York: Routledge, 2010.

Alexievich, Svetlana. *Secondhand Time: The Last of the Soviets*. New York: Random House, 2016.

———. *The Unwomanly Face of War: An Oral History of Women in World War II*. Translated by Richard Pevear and Larissa Volokhonsky. New York: Random House, 2017.

———. *Voices from Chernobyl*. Translated and with a preface by Keith Gessen. Normal: Dalkey Archive Press, 2005.

Bertaux, Daniel, Paul Thompson, and Anna Rotkirch, eds. *Living through the Soviet System*. New Brunswick: Transaction Publishers, 2005.

Cohen, Laurie R. *Smolensk under the Nazis: Everyday Life in Occupied Russia*. Rochester: Rochester University Press, 2013.

Engel, Barbara Alpern and Anastasia Posadskaya-Vanderbeck, eds. *A Revolution of Their Own: Voices of Women in Soviet History*. Boulder: Westview Press, 1998.

Figes, Orlando. *The Whisperers: Private Life in Stalin's Russia*. New York: Metropolitan Books, 2007.

Gheith, Jehanne M. and Katherine R. Jolluck, *Gulag Voices: Oral Histories of Soviet Incarceration and Exile*. New York: Palgrave Macmillan, 2011.

Gildea, Robert, James Mark, and Anette Warring. *Europe's 1968: Voices of Revolt*. Oxford: Oxford University Press, 2013.

Ilic, Melanie and Dalia Leinarte. *The Soviet Past in the Post-Socialist Present: Methodology and Ethics in Russian, Baltic, and Central European Oral History and Memory Studies*. New York: Routledge, 2016.

Inkeles, Alex and Raymond Bauer. *The Soviet Citizen: Daily Life in a Totalitarian Society*. New York: Antheneum, 1959; 1968.

Iwry, Samuel and L. J. H. Kelley. *To Wear the Dust of War: From Bialystok to Shanghai to the Promised Land: An Oral History*. New York: Palgrave Macmillan, 2004.

Khanenko-Friesen, Natalia and Gelinada Grinchenko, eds. *Reclaiming the Personal: Oral History in Post-Socialist Europe*. Toronto: University of Toronto Press, 2015.

Khubova, Daria, Andrei Ivankiev, and Tonia Sharova, "After Glasnost: Oral History in the Soviet Union," in Luisa Passerini, ed., *International Yearbook of Oral History and Life Stories*, Vol. 1, *Memory and Totalitarianism*. Oxford: Oxford University Press, 1992.

Kozlov, Denis. *The Readers of* Novyi Mir*: Coming to Terms with the Stalinist Past*. Cambridge: Harvard University Press, 2013.

Litvinenko, Olga and James Riordan, eds. *Memories of the Dispossessed: Descendants of Kulak Families Tell Their Stories*. Nottingham: Bramcote Press, 1998.

Lourie, Richard. *Russia Speaks: An Oral History from the Revolution to the Present*. New York: Burlingame Books, 1991.

Lyandres, Semion. *The Fall of Tsarism: Untold Stories of the February 1917 Revolution*. Oxford: Oxford University Press, 2013.

Merridale, Catherine. *Night of Stone: Death and Memory in Twentieth-Century Russia*. London: Viking, 2000.

Merridale, Catherine. *Ivan's War: Life and Death in the Red Army, 1939–1945*. New York: Metropolitan Books, 2006.

Messana, Paola. *Soviet Communal Living: An Oral History of the Kommunalka*. New York: Palgrave Macmillan, 2011.

Raleigh, Donald J. *Russia's Sputnik Generation: Soviet Baby Boomers Talk about Their Lives*. Bloomington; Indianapolis: Indiana University Press, 2006.

———. *Soviet Baby Boomers: An Oral History of Russia's Cold War Generation*. New York: Oxford University Press, 2011.

Ritchie, Donald A. *Doing Oral History*. Third Edition. New York: Oxford University Press, 2015.

———. ed., *The Oxford Handbook of Oral History*. New York: Oxford University Press, 2012.

Rogacheva, Maria A. *The Private World of Soviet Scientists from Stalin to Gorbachev*. Cambridge: Cambridge University Press, 2017.

Shapiro, Susan and Ronald Shapiro. *The Curtain Rises: Oral Histories of the Fall of Communism in Eastern Europe*. Jefferson, NC: McFarland and Company, 2004.

Shternshis, Anna. *When Sonia Met Boris: An Oral History of Jewish Life under Stalin*. New York: Oxford University Press, 2017.

Thompson, Paul and Joanna Bornat. *The Voice of the Past: Oral History*. Fourth Edition. New York: Oxford University Press, 2017.

Todorova, Maria, Augusta Dimou, and Stefan Troebst, eds. *Remembering Communism: Private and Public Recollections of Lived Experience in Southeast Europe*. Budapest, Hungary; New York, NY: Central European University Press, 2014.

Tromly, Benjamin. *Making the Soviet Intelligentsia: Universities and Intellectual Life under Stalin and Khrushchev*. Cambridge: Cambridge University Press, 2014.

Vanek, Miroslav and Pavel Mücker, eds. *Velvet Revolution: An Oral History of Czech Society*. New York: Oxford University Press, 2016.

Walke, Anika. *Pioneers and Partisans: An Oral History of Nazi Genocide in Belorussia*. New York: Oxford University Press, 2015.

Zubok, Vladislav. *Zhivago's Children: The Last Russian Intelligentsia*. Cambridge, MA: The Belknap Press of Harvard University Press, 2009.

Index

About the Author

Maria A. Rogacheva earned her PhD in history from the University of Notre Dame. She has taught at the College of William and Mary, Notre Dame, and Beloit College, and currently works at George Mason University. She is the author of *The Private World of Soviet Scientists from Stalin to Gorbachev* (2017).

www.ingramcontent.com/pod-product-compliance
Ingram Content Group UK Ltd.
Pitfield, Milton Keynes, MK11 3LW, UK
UKHW041629031026
15587UKWH00001B/154

9 781498 574341